STATISTICAL GRAPHICS

A WILEY-INTERSCIENCE PUBLICATION

John Wiley & Sons / New York / Chichester / Brisbane / Toronto / Singapore

STATISTICAL

GRAPHICS

Design Principles and Practices

CALVIN F. SCHMID

PROFESSOR EMERITUS
UNIVERSITY OF WASHINGTON

FORMERLY DIRECTOR OF THE CENTER FOR
STUDIES IN DEMOGRAPHY AND ECOLOGY, AND
EXECUTIVE SECRETARY OF THE
WASHINGTON STATE CENSUS BOARD

Library of Congress Cataloging in Publication Data:

Schmid, Calvin Fisher, 1901–
  Statistical graphics.

  "A Wiley-Interscience publication."
  Includes index.
  1. Statistics—Graphic methods. I. Title.

QA276.3.S35   1983   001.4′226   82-19971
ISBN 0-471-87525-2

Printed in the United States of America

10 9 8 7 6 5 4 3 2 1

PREFACE

Statistical charts are a powerful form of visual communication, but in order to achieve their full potential it is essential that the highest standards of quality be observed and rigorously maintained. Unfortunately, a large proportion of the charts produced at the present time fall short of these standards. Descriptively, they represent varying degrees of mediocrity and inferiority. Although this state of affairs is a complicated one, the most crucially important aspect of chart quality inevitably leads to the problem of chart design. Without a doubt, there is a close link between expertise in chart design and chart quality. The need for training and experience in chart design on the part of those responsible for the production of charts is clearly evident. At the present time, however, judged even by the most minimal standards, there are extremely few adequate courses or special training programs in chart design. Moreover, adequate and authoritative treatises or other sources devoted exclusively to the subject of chart design are nonexistent.

With these facts in mind, the present book was planned to fulfill, at least in part, these important needs. As the title implies, this book is concerned with the issues, problems, principles, and practices of chart design. In this respect, it is both complementary and supplementary to *our Handbook of Graphic Presentation.* While the basic emphasis of the *Handbook* is didactic, with specific instructions concerning the design and execution of a wide variety of graphic forms, the present book focuses most of its attention on wider ranging discussions of issues, problems, critiques, pitfalls, standards, innovations, solutions, principles, and practices that relate in one way or another to chart design. However, the importance of didactic objectives has by no means been neglected.

As a prerequisite to clear understanding of the contents of this book, it is essential to recognize the difference between two of the major functions or purposes of statistical charts. In the construction and utilization of statistical charts, emphasis is placed either on presentation or analysis. This distinction is more than conceptual; it is real, but there are instances, of course, where the two functions may overlap and a distinction between them may not be clear. Traditionally, presentation has been the primary function of statistical charts, and today it is still the prevalent one. However, with the rapid development and application of electronic computers in recent years, the analytic function has increased in importance. In order to obviate any misunderstanding, it should be stated unequivocally that the focus of both the *Handbook* and the present work is on the presentational function of statistical charts.

Statistical graphics is not an exercise in draftsmanship or artistry or in the manipulation of electronic computers. To be sure, the mechanics of chart construction are important, but little can be achieved in a qualitative sense without genuine expertise in chart design. The most accomplished draftsmen or artists or the most knowledgeable computer technologists who lack special training and experience in the principles and practice of chart design are simply not competent to produce statistical charts of high quality.

Paller, Szoka, and Nelson, in their recently published manual on computer graphics, *Choosing the Right Chart: A Comprehensive Guide for Computer Graphics Users*, made a significant commentary concerning some of the consequences of incompetence and dilettantism in chart design:

> Computer graphics hardware and software tools are now becoming widely available in business and government. But most computer generated graphics tend to be poorly designed and fail to improve insight or communications. This poor design is a direct result of the fact that most computer graphics are prepared by programmers, researchers, managers, or clerical staff, with little or no formal training in graphic design. When faced with a request to "make a chart," they often use the "default" styles and designs available from computer graphics packages. But many of these "defaults" were also developed by programmers who had no graphics design experience.

In recent years, the electronic computer has served as a significant catalyst in the development of a renascent interest in statistical graphics as well as facilitating an unprecedented production of statistical charts, and its potential future role in statistical graphics will be far greater than it has been in the past.

It is important not to lose sight of the fact that the mechanical aspect of chart production, though recognizably important, represents a very different dimension from chart design. Whether a chart in its final form is executed manually or by computer, its basic quality and effectiveness as a communication device are determined by the person responsible for its design.

In evaluating the effectiveness and quality of statistical charts, consideration should be given to their role in the broader communication process. In its simplest form, the communication process consists of three major interrelated elements: the chartmaker, the chart itself, and the chart user. The effectiveness of a chart in the communication process depends on its quality as well as on the experience, intelligence, and visual perception of the chart user. The closer the link that exists between the person responsible for the design and execution of the chart and the chart user, the greater the quality and quantity of information that the user is able to derive from it. Moreover, it is obvious that the communication process can be severely flawed or completely destroyed by the production of inferior charts or by the ignorance, inexperience, or other limitations of the chart user. Graphic specialists are well aware of the reciprocal relationship among the elements in the communication process, but, of course, circumstances demand that their attention be focused mainly on statistical charts and their role as integral parts of a communication system.

A special presentation feature of this book is the liberal use of the case method. Many charts representing actual cases have been taken from various publications to demonstrate common errors and pitfalls, as well as to show how charts may be redesigned to serve more effectively the purpose for which they were originally intended. In addition to the selection of inferior or poorly designed charts, various superior and exemplary charts are included to illustrate special or unusual applications, qualities, or standards.

Much can be learned about design principles and standards from the study of both well-designed and poorly designed charts. Many of the charts that have been selected are of necessity subjected to fairly rigorous evaluation and not infrequently to adverse criticism. However, every effort has been made to be objective and avoid any semblance of a captious, dogmatic, or doctrinaire attitude. Clearly defined and generally accepted criteria based on long and well-established principles, standards, and practices in graphic presentation have been carefully observed in evaluating the charts. Wherever relevant and appropriate, especially in the case of substandard charts, specific errors and deficiencies are clearly indicated, described, and interpreted within the context of basic design standards and principles. The success of an illustrative and didactic technique of this kind depends on the integrity and thoroughness with which such a procedure is followed. The importance of the numerous charts that have been selected as cases for critical evaluation and illustration is fully recognized, and, of course, they represent a valuable contribution.

I am grateful to the many individuals and organizations for permission to reproduce a large proportion of the charts included in this book. The reader will find that the source along with one or more explanatory notations is shown for every chart. Appreciation is extended to the Graduate School of the University of Washington for a grant that defrayed part of the cost incurred for editorial and drafting services. I am under special obligation to Vincent A. Miller, Department of Sociology, University of Washington, for editorial and statistical assistance, and I am indebted to Stanton E. Schmid, Vice President for University Relations, Washington State University, and John C. Sherman, Professor of Geography, University of Washington, for their critical reading of the manuscript and for offering valuable suggestions. Credit is extended to Vivian Lomax and Debra Slotvig for their drafting assistance.

CALVIN F. SCHMID

Seattle, Washington
October 1982

CONTENTS

STATISTICAL GRAPHICS

CHAPTER ONE

INTRODUCTION

Basic Issues, Problems, Principles, and Practices of Chart Design

ALTHOUGH THE FIELD OF STATISTICAL graphics represents a powerful system of visual communication, it has been burdened for many decades with a substantial number of questionable practices and unresolved problems. These practices and problems are frequently a source of confusion and uncertainty to those interested in the design and construction of statistical charts, as well as posing a serious obstacle in the development of the discipline itself. Moreover, these problems have not been given the serious consideration they deserve. Traditionally, their recognition has been largely in the form of superficial comments and caveats. However, in recent years many techniques and principles of graphic presentation have been subjected to experimental examination, especially by cartographers, psychophysicists, and psychologists. As pioneering efforts, many of these empirical studies represent constructive contributions toward a better understanding of graphic techniques, while others, largely because of methodological limitations are of little significance. Generally, in a fundamental sense, these analyses have barely touched most of the issues and problems of statistical graphics. As far as the future is concerned, the role of empirical analyses in the development of statistical graphics should become increasingly important.

It must not be assumed, of course, that over the years the discipline of statistical graphics has made little or no progress, or that statistical graphics is unique in the prevalence and complexity of its problems.

Perhaps statistical graphics is different from many other disciplines in that no general survey has ever been made of its basic problems along with questionable or uncertain principles and practices. This book has been designed to fulfill this need at least in a preliminary fashion. Without being overly pretentious, one of its goals represents a stocktaking enterprise. It is a systematic effort to summarize and analyze a number of significant issues and problems in order to clarify as well as resolve some of them.

In the analysis and presentation of a number of issues and problems in this book, a considerable amount of theoretical and empirical data has been derived from cartography, psychophysics, and psychology. Particularly during the 1960s and 1970s, the cartographer has devoted an extraordinary amount of effort to the principles and practices of map design, many of which are not only relevant but directly applicable to statistical graphics. Paradoxically, during this same period of time and even longer, statistical graphics as an area of research has been virtually ignored. Important as the contributions of cartography, psychophysics, and psychology have been in recent years, it probably will be several more decades before answers based on empirical study can be found for many of the questions that have been raised in this book. In the meantime, it is hoped that the current pace of research in cartography will continue and also that serious attention will be paid to the need for a program of basic research in the field of statistical graphics.

NEED FOR HIGHER STANDARDS IN STATISTICAL GRAPHICS

Certainly, one of the major objectives in preparing a book of this kind is to create an awareness among chartmakers and chart users of the shortcomings, needs, problems and possible future developments of statistical graphics. In a sense, it is a brief inventory of the status and potentials of the discipline. Eventually, perhaps a step such as that represented by the present undertaking will help to raise professional standards as well as improve the quality and effectiveness of statistical charts, especially as media of visual communication as well as analytical tools. Because of the superabundance of evidence in the form of poorly designed, misleading, and/or dilletantish charts in newspapers, magazines, and even "learned" books and journals, it requires little effort to convince anyone of the need to improve the quality of statistical charts. Also in this connection, where professional chartmakers are involved, improved training, higher performance standards, and the eventual achievement of genuine expertise would add much to the quality of charts that are produced. However, it must be recognized that there is the anomalous and not uncommon practice in some organizations of relegating chartmaking to lower echelon office workers, artists, and others who lack training in statistical graphics as well as other essential knowledge and skills.

The following excerpts from a letter written by the author a few years ago to the editor of one of the journals in the social sciences illustrate the low quality of statistical graphics found even in scientific publications. Strange as it may seem, this journal is the official organ of a discipline in which close to 100 percent of the membership has been exposed to at least one (for some members, perhaps as many as four or five) college or university courses in statistics and research methodology, to say nothing of additional courses in mathematics. Furthermore, the substantive content and methodology of the field are in one form or another characteristically and overwhelmingly statistical. Quoting from the letter:

I should like to suggest that the professional value and character of (X)-(name of Journal) would be substantially enhanced if more serious consideration was given to the quality of graphic material. In making this statement, I do not imply that (X) has lower standards of graphic presentation than other journals in the social sciences. . . .

It is indeed strange—perhaps even incredible—that social scientists and statisticians who would be outraged by ungrammatical sentences, misspelled words, or even improper punctuation, are so indifferent, if not actually receptive, to crude, clumsy, and amateurish charts. Without much doubt, this state of affairs reflects the extraordinary prevalence of graphic illiteracy among practitioners in these disciplines.

In the most recent issue of (X), there is a total of 16 charts. Judged by any reasonable standards, only 3 could be considered "acceptable," while the remaining 13 could be characterized as ranging from "inferior" to "egregious."'. . .

I do trust, however, that immediate as well as careful attention will be given to the problem of developing higher standards of graphic presentation for (X).

To imply that the ancient Chinese proverb "One picture is worth ten thousand words" is generally applicable to statistical charts produced at the present time is certainly not pertinent nor realistic. However, K. W. Haemer suggests that, providing the chart is accurate, relatively simple, easily interpretable, forceful, convincing, revealing, and attractive, the proverb is relevant and meaningful when applied to statistical charts: "One picture *can be* worth *a thousand* words *or figures.*"

Well-designed statistical charts have certain advantages over tabular, textual, and other numerical and verbal techniques. First, they communicate quickly and directly: They not only save time, but allow the chart user to grasp the main features and implications of a body of data at a glance. Second, they are forceful by providing emphasis to a complete, coherent, and decisive message. Third, they are more convincing: They do more than merely state a point; they actually demonstrate it. Fourth, they are more revealing: They can readily clarify data, frequently bring out hidden facts and relationships, and stimulate as well as aid, analytical thinking and investigation. Fifth, because of their appearance, they attract attention and hold the reader's interest, thus making the data more inviting and provocative.[1]

FUNDAMENTAL THEME AND EMPHASIS OF THIS BOOK: IMPORTANCE OF PRINCIPLES AND PRACTICE OF CHART DESIGN

In presenting the many problems and issues pertaining to statistical graphics, the basic theme of every chapter is the fundamental importance of the principles and practice of chart design. The development of statistical graphics as a sound, viable discipline, including competence among practitioners in this field, is ultimately linked with the development and general acceptance of a body of clear, meaningful, reliable,

---

[1] K. W. Haemer, *Making the Most of Charts*, New York: American Telephone and Telegraph Company, 1960, p. 3.

and workable principles and techniques of chart design in the form of tested expertise and realistic and carefully formulated standards. Also, the multidisciplinary as well as interdisciplinary facts and implications of statistical graphics are discussed. Statistical graphics cuts across many disciplines in which it plays a vital role, for example, statistics, cartography, computer science, engineering, communication, economics, and demography. Also, in a reciprocal sense it can readily be observed that statistical graphics has been nourished by contributions from a number of other disciplines, including statistics, psychology, psychophysics, cartography, graphic arts, engineering, and computer science.

The term "design" as applied to statistical graphics is a systematic process embodying the conception of ideas and the creation of tangible plans directed toward the solution of specific problems relating to the visual communication of statistical information. The primary product of the design process is the creation of a model or schema in the form of a well-ordered and efficient graphic structure for the purpose of transmitting by visual communication a body of statistical data. In order to achieve the most successful design, it is essential to take into consideration the many conditions and factors that affect the communication process such as user requirements, user skills, circumstances of use, type, and complexity of information, and technical problems pertaining to chart construction. Decisions must also be made concerning the kinds and placement of individual symbols including the ordering and proper assignment of elements into visual levels of significance. These and other related questions will be discussed in more detail later in the book.[2]

TWO PRIMARY FUNCTIONS OF STATISTICAL CHARTS: PRESENTATION AND ANALYSIS

In order to understand more fully the basic problems and issues covered in this book, it is essential to recognize that statistical charts can perform two major functions: presentation and analysis. As far as the application of these respective functions is concerned, the distinction is one based both on purpose and on emphasis. In actual practice there may be times when the two functions of analysis and presentation are so inextricably interrelated that differentiation is difficult, if not impossible. With an ever increasing development of computer technology, combining or integrating analysis and display (presentation) has been been greatly facilitated. However, where the purpose of a chart is to illustrate, describe, elucidate, interpret, and transmit information, its primary function is presentation. When there is a specific application of graphic techniques to a well-defined statistical problem where exploration, measurement, calculation, and the derivation of relationships are basic objectives, the analytical function is clearly indicated. Particularly in recent years, the importance of statistical charts as analytical tools is attested to by numerous publications—both books and journal articles—on this subject.[3] Furthermore, whether statistical charts are used for analysis or presentation they serve as essential media of visual communication. It can be correctly stated, however, that the role of communication is more important when the function of a chart is primarily that of presentation. In discussing the many problems and issues included in this book the orientation and emphasis are frankly focused on presentation. In a real sense this book has been designed as a sequel or supplement to our *Handbook of Graphic Presentation*.[4] As the title indicates, this book represents a manual pertaining to the design and construction of statistical charts with special emphasis on presentation.

In light of the foregoing discussion, a clear understanding of the two major functions of statistical charts—analysis and presentation—may seem obvi-

---

[2] For a comprehensive analytical and theoretical system of rules, forms, signs, symbols, principles, and structures of graphic displays, see Jacques Bertin, *Semiologie Graphique: les diagrammes, les reseaux, les cartes,* Paris-La Haye: Mouton; Paris: Gauthier-Villars, 1967 and 1973. Also see Jacques Bertin, *Graphics and Graphic Information Processing,* Berlin and New York: Walter De Gruyter, 1981, passim; J. S. Keates, *Cartographic Design and Production,* New York: Longman, 1973, pp. 29–33; Borden D. Dent, "Simplifying Thematic Maps Through Effective Design: Some Postulates for the Measurement of Success, *Proceedings of the American Congress on Surveying and Mapping,* Fall Convention 1973, pp. 243–251; Alan DeLucia, "Design: The Fundamental Cartographic Process," *Proceedings, Association of American Geographers,* **6** (1974), 83–87; Josef F. Blumrich, "Design," *Science,* **168** (1970), 1551–1554.

---

[3] An up-to-date review of graphical analytical techniques including an extensive bibliography will be found in Howard Wainer and David Thissen, "Graphical Data Analysis," *Annual Review of Pyschology,* **32** (1981), 191–241.

[4] Calvin F. Schmid and Stanton E. Schmid, *Handbook of Graphic Presentation,* New York: John Wiley & Sons, 1979.

ous, but, paradoxically, there appears to be a certain amount of deep-seated and widespread confusion concerning these facts. This confusion seems to be particularly common among those—mostly statisticians and computer specialists—who only recently have acquired a professional interest in statistical graphics. Perhaps their lack of background, experience, and expertise in statistical graphics would account for their failure to understand this seemingly elementary question. In any case, a special effort has been made to clarify the basic distinction between analysis and presentation since it is crucial to an understanding of the problems and issues presented in this book.

Traditionally, presentation has always played a dominant role in statistical graphics, and this role has continued up to the present time. Almost 200 years of distinguished history bears witness to this fact. Beginning with the publication of the first edition of *The Commercial and Political Atlas* by William Playfair—the "father" of modern charting techniques—in 1786, followed by innumerable contributions through the decades, the predominant emphasis on presentation in statistical graphics is very apparent. To cite only a few instances that substantiate this fact, H. G. Funkhouser's notable history of the discipline, published in 1937, "Historical Development of the Graphical Presentation of Statistical Data," a more recent (1962) historical study by Paul J. FitzPatrick, "The Development of Graphic Presentation of Statistical Data in the United States," and Willard C. Brinton's landmark contribution, *Graphic Methods for Presenting Facts* (1914) are devoted almost exclusively to the presentation function of statistical charts. The most recent revision of *Time-Series Charts* (1979, p. ix) by the National Standards Committee Y15, Preferred Practice for the Presentation of Graphs, Charts and Other Technical Illustrations states that

> This standard is not concerned with the analysis of time series; therefore, it is not directly concerned with analytical time-series charts. Some of the principles and many of the procedures listed apply to analytical charts as well as to presentation charts, but the emphasis throughout this standard is on charts for presentation.[5]

Additional evidence of the dominant importance of the presentation and communication aspects of statistical graphics is revealed in a recent survey by Stephen E. Fienberg and his associates. The survey rep-

resents a content analysis of the *Journal of the American Statistical Association* and of *Biometrika,* covering a period from 1921 to 1975, in order to ascertain the extent, trends, and purpose of graphics utilized by statisticians.[6] The basic data in this survey are classified according to the amount of space devoted to the following threefold uses: (1) graphs and charts for displaying data and illustrating the results of analysis; (2) graphs and charts for analytical purposes, and (3) plots and graphs with elements of both display and analysis. Two major conclusions derived from this study indicate that: (1) Since 1921 there is clearly a "decline in the use of statistical graphs . . . at least within two of our major statistical journals." (2) The predominant type of utilization of graphics for both journals during the entire period was for display and communication, not analysis.

Again, it must be emphasized that these conclusions are based on only two journals exclusively devoted to the subject of statistics. A question that might be asked is: If similar data were available for the hundreds of journals in the social sciences, education, engineering, physical sciences, cartography, geography, medicine, journalism, and other fields, would the results be substantially different? Of course, any answer would be largely conjectural or impressionistic. Nevertheless, it is conceivable that there has been an upward trend in the utilization of statistical graphics in a number of fields and specialities other than statistics and that the utilization of statistical graphics has been predominantly for illustrative and communication purposes.

---

[5] More complete citations of publications in this paragraph are as follows: William Playfair, *The Commercial and Political Atlas* (3rd ed.), London: J. Wallis, 1801; H. G. Funkhouser, "Historical Development of the Graphical Representation of Statistical Data," *Osiris,* **3** (1937), 369–404; Paul J. FitzPatrick, "The Development of Graphic Presentation of Statistical Data in the United States," *Social Science,* **37** (October 1962), 203–214; Willard C. Brinton, *Graphic Methods for Presenting Facts,* New York: The Engineering Magazine Co., 1915; American National Standards Committee Y15, *Time-Series Charts,* New York: The American Society of Mechanical Engineers, 1979.

[6] Stephen E. Fienberg, "Graphical Methods in Statistics," *The American Statistician,* **33** (November 1979), 165–178. Also, see Vincent P. Barabba, "The Revolution in Graphic Technology," in *Proceedings of the First General Conference on Social Graphics,* 1978, Technical Paper No. 49, United States Bureau of the Census, pp. 1–21.

WHY THE DIMINISHED IMPORTANCE OF GRAPHICS AMONG PROFESSIONAL STATISTICIANS

The decline in the utilization of graphics by statisticians during the 1921–1975 period is linked to certain historical trends and changes occurring within the science of statistics. These changes reflect certain extraordinary shifts in emphasis and orientation. If one were required to point to a single index or event that presaged these radical changes in statistics during this period, the inevitable selection would be the publication in 1925 of Ronald A. Fisher's *Statistical Methods for Research Workers*. A new era was ushered in where analytical and inductive methodology superseded the more traditional approach. To be sure, other mathematicians and statisticians before and after Fisher have made important contributions to present-day statistical methodology, but it was Fisher who combined into one powerful tool the design of experiments, exact tests of significance, theory of estimation, and simple but adequate arithmetical processes required in the interpretation of experimental data.

As the number of statisticians trained in "modern" statistics increased, graphic presentation came to be regarded as more and more irrelevant to what was considered to be the proper domain of statistics. As time went on, this became the predominant point of view and was reflected in textbooks, statistical journals, and courses in statistics. For example, most pre-1930 textbooks in statistics devoted at least one and not infrequently two chapters to graphic presentation. By the end of World War II, a large proportion of texts treated statistical graphics in a cursory fashion or eliminated the subject altogether. To be sure, there were some, in certain applied areas such as business and the social sciences, that included discussions of statistical graphics for pedagogical, vocational, or other reasons. As statistical science has evolved since 1930 or so it seems only logical that the role of graphics would recede in importance or actually become irrelevant except as an ancillary tool of analysis or an incidental technique of visual communication.[7]

Almost 60 years ago, Karl G. Karsten made the following statement: "Chart-making is an art which all can practice. But there will always be a world of difference between the charts of amateurs and those of master-statisticians."[8]

Today such a statement sounds anachronistic. Perhaps it could be said that it contains a modicum of truth, but an authentic, meaningful, and modernized version would require substantial qualification and elucidation. To be sure, there is a wide gap between the quality of charts produced by the amateur and dilettante and the quality of charts produced by the graphic specialist. But the contemporary graphic specialist may not necessarily be a "master statistician." He or she is just as likely to be a cartographer, computer specialist, social scientist, designer, communications expert, or engineer. Up until the 1930s statisticians dominated the field of graphic presentation. Virtually all statisticians had some training in statistical graphics, but at the present time this is not the case. With these facts in mind, the present writer made the following comments a few years ago:

In recent decades, statisticians by and large have become indifferent and neglectful of graphic presentation. As far as the mainstream of contemporary statistics is concerned, graphic presentation has been shunted into a marginal niche. This trend is clearly reflected in the programs of statistical societies, in statistical journals, in university courses in statistics, and in treatises on statistics. . . . One observes that with this growing indifference and neglect, the incidence of graphic illiteracy . . . among statisticians seems to have increased. This observation can be substantiated by the clumsy and amateurish charts produced by or under the direction of statisticians.[9]

RENAISSANCE OF STATISTICAL GRAPHICS

Significantly, since the early 1970s, there have been clear indications of a renascent interest in graphic

---

[7] Calvin F. Schmid, "Contributions of an Obscure and Neglected Period in the History of Statistical Graphics: 1930 to 1960," First General Conference on Social Graphics, October 1978 (in press).

[8] Karl G. Karsten, *Charts and Graphs,* New York: Prentice-Hall, 1923, p. x.

[9] Calvin F. Schmid, "The Role of Standards in Graphic Presentation," American Statistical Association, *Proceedings of the Social Statistics Section: 1976,* Part I, pp. 74–81. Reprinted in United States Bureau of the Census, *Graphic Presentation of Statistical Information,* Technical Paper 13 (1978), pp. 69–78.

presentation among many statisticians. This trend is attested to by the reorganization of the Committee on Statistical Graphics and by special programs of the American Statistical Association; by programs of the International Statistical Institute; by the newly organized Council on Social Graphics, and the first General Conference held under its auspices; and by the growing interest in computer graphics as well as various research projects and numerous publications on the subject of statistical graphics. Presumably, the pendulum is now beginning to swing in the other direction. Today the climate is more conducive to the progressive development of graphic presentation in terms of higher standards, innovation, improved techniques, better trained specialists, and wider usage and acceptability than it has been in several decades. In the paper cited previously, Stephen E. Fienberg makes an outspoken and effective plea to statisticians:

> Clearly, one of the things we need in the area of statistical graphics is more.... First, we must teach statisticians and others how and when to draw good graphic displays of data. Second we must encourage them to use graphical methods in their work and in the material they prepare for publication. Third, we must change the policies in our professional journals so that graphics are encouraged, not discouraged.[10]

Although a substantial knowledge of statistics is an essential ingredient for expertise in graphic presentation, graphic presentation can no longer be identified exclusively with statistics or considered an integral branch or specialty of statistics. As graphics presentation has evolved, it has been fed by a number of interdisciplinary streams. In addition to statistics, there have been important elements derived from (1) technical drawing, (2) graphic arts, (3) cartography, (4) psychology (5) psychophysics, (6) computer science, and (7) communication theory and practice.

STATISTICAL GRAPHICS: AN EMBRYONIC DISCIPLINE?

Perhaps in time, it is conceivable that statistical graphics will evolve into an autonomous discipline or profession with a well-established theoretical, practical, and organizational structure.

---

[10] Stephen E. Fienberg, "Graphical Methods in Statistics," *The American Statistician*, **33** (November 1979), 165–178.

The preceding facts have important implications for both the vitality and future development of statistical graphics and the professional training and standards of the individual specialist who aspires to genuine proficiency and expertise in statistical graphics. For example, as far as the individual specialist is concerned, it is obvious that expertise in statistical graphics includes much more than a knowledge of statistics per se, skill in drafting, a thorough understanding of computer technology, or proficiency merely in one or two of the other disciplines that have been mentioned. Rather, to a greater or lesser degree, it involves all of them. It is only through the acquisition of certain principles and techniques derived from these disciplines, along with the necessary skills in applying them, that genuine expertise in statistical graphics is possible.

PROFICIENCY IN CHART DESIGN: AN ALL IMPORTANT REQUISITE FOR EXPERTISE IN STATISTICAL GRAPHICS

The most fundamental requisite for proficiency in statistical graphics is a thorough grasp of the principles and practice of chart design including all types of charts, ranging from simple arithmetic line graphs to more elaborate statistical maps and three-dimensional stereograms. This idea is elucidated in a recent paper by George F. Jenks. Although, as a cartographer, Jenks had statistical maps in mind, the implications of the following statement apply to all other forms of statistical charts.

> Statistical maps are now produced in greater numbers and by persons in a wider range of disciplines than ever before. It is unfortunate that this increased production has not been accompanied by an improvement in map quality. All too frequently, contemporary mapmakers evidence a lack of understanding of the primary function of statistical maps, the symbolic language of mapping, and the effect that data manipulation plays in map communication. Coupled with these deficiencies, there is a general lack of appreciation of the basic elements of graphic design. Furthermore, the expanded use of the computer in mapmaking seems to be related to and may even foster low quality since programmers are generally untutored in cartographic design and communication.[11]

What Jenks says about the low quality of a substantial proportion of maps produced at the present time is also unmistakably relevant to statistical charts. Poorly designed and executed statistical charts may be made using either computers or manual techniques. Generally, substandard charts reflect inadequate knowledge and skill in chart design. There are instances, of course, where the computers and auxiliary equipment used are not adapted to the construction of high-grade charts. In this connection it should never be assumed that elaborate and costly computers with "raster scans, refresh tubes and output devices of modern technology" will provide incontestable insurance against the production of inferior charts. Facts clearly indicate that there is no paucity of examples of poorly prepared charts produced on the most up-to-date computers with every known technological improvement that may represent an investment of as much as a million dollars.

These comments are not meant to imply in any way a blanket criticism of computer-generated charts. The production of inferior charts lies much deeper than the mere mechanics of their preparation. Rather, the criticism is directed at the inferior design skills and standards of those responsible for their production. It is obvious that computers are an indispensable adjunct to the production of certain types of statistical charts, especially those constructed for analytical purposes. Furthermore, as far as the future is concerned, computers will play an ever increasing role in statistical graphics.

The charting process involves two interrelated dimensions: The first and more demanding is chart design; the second is concerned with the mechanics of chart construction. In chart design, emphasis is placed on creativity, ingenuity, and ideation—the reduction of ideas to readily interpretable and meaningful visual form. The mechanical aspects of the charting process involve manual and other skills, such as drafting or proficiency in operating an electronic computer. There are extremely rare cases where one person possesses both extraordinary expertise in design and outstanding skill in the mechanics of chart construction. However, it seems self-evident that chart design is a body of knowledge and expertise distinct from drafting skills per se or the ability to operate an electronic computer. The most skilled draftsmen or the most experienced computer

specialists, to say nothing of their run-of-the-mill counterparts, who do not possess extensive and well-rounded background and training in graphic design and supporting disciplines, are simply not competent to prepare superior statistical charts. Perhaps the most significant concern in the field of graphics presentation is the production of statistical charts of highest quality—charts that are effective, meaningful, attractive, and reliable. This means higher standards of professional competence, which in turn depend on the chartmaker's training and experience in graphic design. Although recognizably important, the mechanics of charts—whether manual or automated—represent a different dimension.[12]

CONTRIBUTIONS OF CARTOGRAPHY TO THE THEORY AND PRACTICE OF CHART DESIGN

In order to obtain deeper insight into certain basic theoretical and practical problems of chart design, much can be learned from the experience and research of the cartographer. As indicated previously, statistical graphics and certain aspects of cartography are closely related, if not actually identical. Both cartography and statistical graphics represent communication systems based on graphic symbols and both are involved in the design and construction of statistical charts. In cartography, the basic referent is space and the all important vehicle is the map, whereas in statistical graphics a greater variety of graphic forms is customarily used. It has been estimated that "thematic cartography accounts for far more than 80% of the map production."[13]

Assuming that this figure is approximately correct, the implications of this statement are self-evident. Since a very substantial proportion of "thematic maps" can be classified as "statistical maps," it is obvious that there is a broad, significant, and overlapping area between cartography and statistical graphics.

[11] George F. Jenks, "Contemporary Statistical Maps—Evidence of Spatial and Graphic Ignorance," *The American Cartographer*, **3** (1976), 11–19.

[12] Calvin F. Schmid and Stanton E. Schmid, *Handbook of Graphic Presentation*, New York: John Wiley & Sons, 1979, p. 277.

[13] Karl-Heinz Meine, "Cartographic Communication Links and a Cartographic Alphabet," in Leonard Guelke (ed.), *The Nature of Cartographic Communication*, Monograph No. 19, *Cartographica*, Toronto: York University Press, 1977, pp. 72–91.

These facts do not necessarily justify the inference that the various theories and techniques developed by the cartographer are applicable literally and in toto to statistical graphics. However, it does mean that those involved in the design and construction of statistical charts can profit immensely from the contributions of the cartographer in the form of basic theory, helpful suggestions, design principles, enlightened guidance, and empirical facts.

A BASIC CARTOGRAPHIC PRINCIPLE: THE MAP, AN ESSENTIAL PART OF A COMMUNICATION SYSTEM

In recent years the cartographer has placed much emphasis on the map as an essential part of a broader communication system and in recognition of this role, has developed a number of models to explain various factors in the communication process. For example, one very popular model is that of A. Kolacny, published in 1968.[14]

Kolacny posits several premises and assumptions that help to clarify certain theoretical and practical implications of his model. For example, in the past, map design and construction were almost exclusively based on decisions made by the cartographer with lit-

tle consideration for the needs and expectations of the map user. Kolacny believes that this state of affairs should be reversed so that the creative work of the cartographer should be based on the needs, interests, and subjective conditions of the map user, that is, on an intimate knowledge of the conditions that constitute the problems associated with the use of maps. Cartography is interested not only in the creation or production of maps, but also in the utilization or consumption of maps—two components of a coherent (and, in a sense, indivisible) process in which cartographical information originates, is communicated, and produces an effect.

According to Kolacny's model, the seven principal factors active in the process of communicating information are:

1. U_1—reality (the universe) as seen by the cartographer.

2. S_1—the cartographer.

3. L—cartographic language as a system of map symbols and rules for their use.

4. M—map.

5. S_2—map user.

6. U_2—reality (universe) as seen by the map user.

7. I_c—cartographic information

The creation and communication of cartographic information is a complex process of activities and operations with feedback circuits on various levels.

CONCEPT OF CARTOGRAPHY AS A COMMUNICATION SYSTEM: ITS INFLUENCE ON MAP DESIGN AND MAP DESIGN RESEARCH

The following quotations indicate the prevalence among contemporary cartographers of the view that cartography is a communication system. Also, it will be found that this concept has had a pervasive influence on the principles and practice of map design as well as on scores of empirical studies pertaining to map design.[15]

---

[14] A. Kolacny, *Cartographic Information—A Fundamental Notion and Term in Modern Cartography,* Prague: Czechoslovak Committee on Cartography, 1968; A. Kolacny, "Cartographic Information—Fundamental Concept and Term in Modern Cartography," *Cartographic Journal,* **6** (1969), 47–49. Indicative of its popularity, the Kolacny model has been re-published in a number of cartographic sources either in its original or modified form. The following references might be cited as examples: Arthur H. Robinson and Barbara Bartz Petchenick, *The Nature of Maps,* Chicago: The University of Chicago Press, 1976, Chapter 2; Leonard Guelke, "Cartographic Communication and Geographic Understanding," in *The Nature of Cartographic Communication,* Leonard Guelke (ed.), Monograph No. 19, *Cartographica,* Toronto: York University Press, 1977, pp. 129–145; K. A. Salichtchev, "Some Reflections on the Subject and Method of Cartography After the Sixth International Cartographic Conference," *Ibid.,* pp. 111–116; Michael Wood, "Human Factors in Cartographic Communication," *The Cartographic Journal,* **9** (1972), 123–132; Joan Roberts, "An Essay in Communication," *Cartography,* **10** (1977), 100–102; David Woodward, "The Study of the History of Cartography: A Suggested Framework," *The American Cartographer,* **1** (1974), 101–115. For a discussion of other models of cartographic communication, see George F. McCleary, "How to Design an Effective Graphics Presentation, Cambridge, Mass.: Harvard Library of Computer Graphics, Laboratory for Computer Graphics and Spatial Analysis, 1981, pp. 15–19.

[15] For example, see a recent paper that discusses the conceptual and methodological relationship between the symbolization process and cartography as a communication system. Mei-Ling Hsu, "The Cartographer's Conceptual Process and Thematic Symbolization," *The American Cartographer,* **6** (1979), 117–127.

All in all, there seems no doubt that the field of cartography has opened wide its arms to welcome the concept that it is a communication system. . . .[16]

. . . the first principal task of cartography is to increase our knowledge of the cartographic communication process. . . .[17]

. . . a major thrust of current cartographic investigation deals with exploring methods of making the map more effective as a *means of communication.*[18]

To say that maps are *the* language of geography is to imply two things. First, they are unsurpassed at present as a two-dimensional vehicle of communication. Second, they are devices to communicate geographical concepts, not only to geographers but to the public at large.[19]

Cartographic communication describes the process whereby information is selected, symbolized on a map, and subsequently perceived, recognized, and interpreted. Communication is complete only when the coded message has been deciphered and interpreted.[20]

It is evident that in order to achieve optimal communicative efficiency in cartography or statistical graphics, well-designed charts are indispensable. Those interested in planning and preparing maps and charts must rely primarily on principles and practices based on tradition, convention, intuition, and personal experience. Statisticians, cartographers, psychophysicists, psychologists, and others have recognized this fact for some time and believe that the solution lies in

experimental research. For example, Bruce Mudgett, a statistician, commented in 1930 that

Very few of the graphic methods in use today have been tested widely by experiment. Their validity is founded largely on logical considerations; and while the importance of this foundation is not to be denied, no less is it to be denied that these conclusions deserve to be subjected to careful testing wherever experimental methods may be applied and where the results may be set forth in quantitative terms.[21]

Recently (1975) another statistician, William H. Kruskal expressed similar ideas concerning statistical graphics.

. . . in choosing, constructing, comparing, and criticizing graphical methods we have little to go on but intuition, rule of thumb, and a kind of master-to-apprentice passing along of information. You need only look at a good text on statistical graphics. Much of its advice will be excellent, no doubt, but it will also be dogmatic or arbitrary, in the sense that there is neither general theory nor systematic body of experiment as a guide. What we have instead are accumulated experiences, social conventions, and prescriptions.[22]

The following extended quotation from a cartographer, Alan De Lucia, clearly expresses this problem as well as the rationale behind design research.

The basic problem in all map design is to logically select and effectively combine a set of graphic symbols in order to display spatially distributed data in the most efficient, easily perceived manner. The question is: What kind of design knowledge is the cartographer going to have to use as a basis for making the decision which will ultimately provide the most visually efficient map for the user?

There are two alternatives. Until recently, the cartographer had to rely primarily upon intuitive design concepts developed either by himself or others over long years of experience and trial and error. . . .

---

[16] Arthur H. Robinson and Barbara Bartz Petchenik, "The Map as a Communication System," *The Cartographic Journal,* **12** (1975), 7–15.

[17] Karl-Heinz Meine, "Cartographic Communication Links and a Cartographic Alphabet," in Leonard Guelke (ed.), *The Nature of Cartographic Communication,* Monograph No. 19, *Cartographica,* Toronto: York University Press, 1977, pp. 72–91.

[18] Borden D. Dent, *Perceptual Organization and Thematic Map Communication: Some Principles for Effective Map Design with Special Emphasis on the Figure-Ground Relationship,* Worcester, Mass.: Cartographic Laboratory, Clark University, Chapter 1.

[19] Christopher Board, "Cartographic Communication and Standardization," *International Yearbook of Cartography,* **13** (1973), 229–236.

[20] Michael Wood, "Human Factors in Cartographic Communication, "*The Cartographic Journal,* **9** (1972), 123–132.

---

[21] Bruce D. Mudgett, *Statistical Tables and Graphs,* Boston: Houghton Mifflin Company, 1930, p. 67.

[22] William H. Kruskal, "Visions of Maps and Graphs," in *Auto-Carto II: Proceedings of the International Symposium on Computer-Assisted Cartography,* Washington: Bureau of the Census, 1975, pp. 27–36.

The second alternative, and the one which is essential and necessary to the further advancement of map design, is the development of scientific design principles. These must be firmly based upon the objective testing of cartographic products utilizing experimental techniques already established by human factors engineers, psychologists and physiologists.[23]

HISTORICAL SUMMARY OF RESEARCH IN CHART DESIGN

Although in recent years cartographers have been responsible for the most notable amount and quality of research in chart design, the pioneers in this field have been mainly statisticians and psychologists. In 1926, Walter Crosby Eells of Whitman College (Walla Walla, Washington) published a paper in the *Journal of the American Statistical Association* on "The Relative Merits of Circles and Bars for Representing Component Parts."[24]

Since this study represents the prototype of dozens of psychophysical studies that were to appear later, the following brief remarks may be of special interest. For many decades, graphic specialists, statisticians, and others generally have held the pie chart in low esteem. Eells cites 11 contemporary published sources to emphasize the point and adds that "It is noticeable that all of the quotations given seem to be purely matters of opinion—none show any experimental basis of fact." Sheets of paper containing 15 subdivided circles and the same number of subdivided bars were submitted to a sample of students for interpretation in order to determine comparative rapidity of judgment as well as popularity and appeal. Eells concluded that "circle diagrams were generally superior, and they should be recommended, not only on account of their popularity and psychological appeal, but also on the basis of scientific accuracy...." Following Eells' study, R. von Huhn and Frederick E. Croxton published critiques that raised serious

doubts about Eell's methodology and conclusions.[25] Subsequently, two additional related papers were published pertaining to a comparative evaluation not only of circles and bars but also of squares and cubes.[26] Another significant pre-World War II contribution to graphic presentation in which experts in optics and psychophysicists collaborated with cartographers on a relatively brief monograph entitled *Notes on Statistical Mapping* was published jointly by the American Geographical Society and the Population Association of America.[27] This monograph includes discussions of such topics as point symbols, choropleth and isopleth maps, and analyses of graded shading and textural patterns.[28]

RESEARCH IN CARTOGRAPHIC DESIGN

Arthur H. Robinson shows that design research in cartography can be characterized by five different approaches and basic methodologies,[29] including:

I. Indirect Research Approach
    (a) Empirical methods
    (b) Adaptation of studies in other fields

II. Direct Research Approach
    (c) Census of user reaction
    (d) Task-oriented research
    (e) Psychophysical research

---

[23] Alan De Lucia, "The Effect of Shaded Relief on Map Information Accessibility," *The Cartographic Journal,* **9** (1972), 14–18.

[24] Walter Crosby Eells, "The Relative Merits of Circles and Bars for Representing Component Parts," *Journal of the American Statistical Association,* **21** (1926), 119–132.

[25] R. von Huhn, "A Discussion of Eells' Experiment," *Journal of the American Statistical Association,* **XXII** (1927), 31–36; Frederick E. Croxton, "Further Studies in the Graphic Use of Circles and Bars: Some Additional Data," ibid., **XXII** (1927), 36–39.

[26] Frederick E. Croxton and Roy E. Stryker, "Bar Charts Versus Circle Diagrams," *Journal of the American Statistical Association,* **XXII** (1927), 473–482; Frederick E. Croxton and Harold Stein, "Graphic Comparisons by Bars, Squares, Circles, and Cubes," *Journal of the American Statistical Association,* **XXVII** (1932), 54–60.

[27] John K. Wright et al., *Notes on Statistical Mapping,* New York and Washington: American Geographical Society and Population Association of America, 1938, passim.

[28] Several additional evaluative and analytical studies of statistical graphics completed prior to 1940 will be found in *How Pictures and Graphs Aid Learning from Print,* Technical Memorandum No. 4 prepared by University of Illinois, Division of Communication for the United States Air Force, December 30, 1952. The emphasis in most of these studies is pedagogical.

[29] Arthur H. Robinson, "Research in Cartographic Design," *The American Cartographer,* **4** (1977), 163–169.

The following is a brief description of each of the five specific categories in the foregoing summary: (a) Empirical methods, the most personal and intuitive, are essentially trial-and-error procedures in which successful solutions to specific problems are accepted while others are discarded. (b) Adaptation of studies in other fields refers to borrowing relevant ideas and techniques from other fields, such as psychology, filmmaking, television, and inductive design. (c) Census of user reactions involves survey techniques based either on questionnaires or personal interviews of users of certain kinds of maps for the purpose of eliciting information helpful in formulating principles of cartographic design. (d) Task-oriented research focuses on the uses of particular kinds of maps through tests and interviews with subjects who use them. The primary objective is to improve particular types of maps by testing alternative designs and resolving specific problems. (e) Psychophysical research is characteristically experimental and, in cartographical design problems, relates specifically to visual physical stimuli and psychical response. This type of research encompasses various kinds of so-called perceptual problems that have been pursued extensively in recent years by cartographers and will be discussed in more detail in the following paragraphs.

As indicated previously, the fundamental concern of psychophysical or perceptual research in cartography is to improve the efficiency of the communication process by producing better maps based on an understanding of how users perceive maps. Scores of perception studies conducted by cartographers and psychologists have been devoted to such problems as two- and three-dimensional graduated point symbols, printed shading screens, curves of the gray spectrum, lettering and typography, color, nonquantitative symbols, class intervals for choropleth maps, dots, lines, bars, and isopleths.[30]

GRAPHICACY, AN INDISPENSABLE SKILL IN VISUAL COMMUNICATION

W. G. V. Balchin and Alice M. Coleman coined an interesting and significant word—"graphicacy"—which connotes an intellectual skill for the communication of relationships that cannot be communicated by word or mathematical notation alone.[31] It is a skill needed by both the person who wishes to transmit the communication as well as the person who wishes to

receive the communication. The type of communication is characteristically visual and is mediated through charts, maps, photographs, and so on. Graphicacy represents one of four basic intellectual skills: literacy, numeracy, articulacy, and graphicacy. Literacy includes the basic skills of reading and writing (i.e., communication with written words). Numeracy expresses the ability to communicate in numbers and other mathematical notation. Articulacy signifies the art of spoken communication. As indicated previously, graphicacy implies the ability to communicate visually.

The authors present cogent arguments for all four skills as indispensable underpinnings in a sound educational program. To be fully effective in achieving their highest level of communication, all four basic skills should be integrated. These skills are not interchangeable nor is one a substitute for any other—they are complementary. Neither words nor numbers nor diagrams are simpler or more complex, superior or inferior. They are only more suitable or less suitable for particular purposes and each may range from the very simple to the highly complex. A well-educated person, of course, should not only be able to read and write with facility, but also be articulate, numerate, and graphicate. A challenge is issued to educators to mold the vague ideas of visual aids at large into a more integrated goal of education, to begin

[30] A brief review of perceptual research in cartography will be found in D. Brandes, "The Present State of Perceptual Research in Cartography," *The Cartographic Journal,* **13** (1976), 172–176. It will be observed that a considerable portion of Chapter 7 of this book is devoted to summaries of perceptual research studies. With respect to individual perceptual studies, only a few examples will be cited: James J. Flannery, "The Relative Effectiveness of Some Common Graduated Point Symbols in the Presentation of Quantitative Data," *Canadian Cartographer,* **8** (1971), 96–109; Hans-Joachim Meihoefer, "The Visual Perception of the Circle in Thematic Maps/Experimental Results," *Canadian Cartographer,* **10** (1973), 63–84; George F. Jenks and Fred C. Caspall, "Error on Choropleth Maps: Definition, Measurement, Reduction," *Annals of the Association of American Geographers,* **61** (1972), 217–244; Paul V. Crawford, "Perception of Graduated Squares as Cartographic Symbols," *Cartographic Journal,* **10** (1973), 85–88; Goesta Ekman and Kenneth Junge, "Psychological Relations in Visual Perception of Length, Area and Volume," *Scandinavian Journal of Psychology,* **2** (1961), 1–10.

[31] W. G. V. Balchin and Alice M. Coleman, "Graphicacy Should be the Fourth Ace in the Pack," *The Cartographer,* **3** (1966), 23–28; W. G. V. Balchin, "Graphicacy," *Geography,* **57** (1972), 185–195.

teaching graphicacy at an early stage, and to give graphicacy its rightful place in education. Furthermore, it is apparent that no matter how superior a chart may be in its design and execution, the communication process is doomed to failure unless the reader is reasonably graphicate.

PRESENT STATUS OF PERCEPTION RESEARCH: EVALUATION AND CRITICISM

However, in spite of the substantial amount of perceptual research in cartography in recent years,

> ... it is already becoming clear that major advances in map design that the new emphasis on map perception seemed to promise have not materialized. Map perception research has, with a few notable exceptions, had a slight impact on map design and production.[32]

Another cartographer observes that

> A considerable amount of perceptual research within the general framework of behavioral psychology has been conducted by cartographers during the last ten or fifteen years. However, as one reviews the findings of this research in connection with problems encountered during the normal process of making maps, it doesn't seem to add up to much. No whole theory or set of principles, greater than the sum of the small component parts, has emerged.[33]

Again, in a recently published book on cartography, the author mentions that perception

> studies which provide a fascinating exploration of how people react to various mapping techniques, have been largely ignored. Most of the information was originally intended to be used to improve map design, but it is inconclusive and hasn't received much acceptance by map makers.[34]

Although perception research has failed to provide solutions and guidance to the overwhelming proportion of basic design problems in cartography, it is patently unrealistic to assume that this could be achieved in less than two decades. Scientific development is a slow, arduous process, going forward by small increments. The testing and retesting of relatively simple hypotheses, their modification and further testing and elaboration, and the correlation and synthesis of verified facts and generalizations are a few of the indispensable steps in the development of a body of reliable and meaningful principles and practices. As Karl Pearson pointed out many years ago, there is no shortcut to scientific "truth." If cartographers and other graphic specialists devote sufficient time and effort, particularly if they collaborate closely with psychologists and psychophysicists during the next few decades, perhaps a body of design principles and practices based on a sound scientific formulation can be created. No doubt, cartographic research of the future will be marked by many innovations, including more theoretically sound and productive orientations and more refined methodologies and research techniques. Substantively, it is to be hoped that research of this kind will be expanded to include statistical graphics in all its manifestations. In time, facts and principles derived from various kinds of scientific experiments will play an increasingly important role, but still there will be problems and procedures that transcend scientific analysis. Furthermore, many present-day conventions, practices, and standards that have evolved through experience and tradition will be confirmed and supported by scientific research. Standards, rules, and conventions will continue to exist, but in the future they will be based more and more on scientific knowledge rather than on intuition, impression, and tradition. They also will be subject to change, but the changes will be determined largely by additional scientific insight and understanding. As cartography and statistical graphics become more mature scientifically, professional training and standards most likely will be more demanding. In the future, genuine competence in cartography and statistical graphics will require the same personal qualities that are essential today: intelligence, experience, imagination, creativity, and sound judgment.

[32] Leonard Guelke, "Perception, Meaning, and Cartographic Design," *Canadian Cartographer,* **16** (1979), 61–69.

[33] Barbara Bartz Petchenik, "Cognition in Cartography," in L. Guelke (ed.) *The Nature of Cartographic Communication,* Cartographica Monograph No. 19 (1977), 117–128.

[34] Philip C. Muehrcke, *Map Use Reading, Analysis and Interpretation,* Madison, Wisconsin: J. P. Publications, 1978, p. viii.

SUGGESTIONS FOR IMPROVING CHART DESIGN PRINCIPLES AND PRACTICES

In light of the foregoing discussion, it is apparent that the context and referents in any attempt to present a summary of the principles and practices of chart design must be based largely on the accumulated precepts, conventions, traditions, techniques, and standards that have evolved over the years from the experience, thinking, and formulations of hundreds of practitioners. As previously indicated, the contributions of experimental studies to design principles and practices are limited in number, scope, and reliability.

Because of restricted space, this discussion must be brief. However, in striving for brevity, this section risks manifesting some of the characteristics of a "do-it-yourself" manual, or a "cookbook" of design recipes, or a simplistic listing of "do's and don'ts." Care has been taken to avoid these pitfalls.

MECHANICS OF PREPARING CHARTS

The mechanics of constructing charts in their final form, whether manually or by electronic computer, is without question one of the most crucial steps in the chart production process. The most impeccably designed chart can be ruined by shoddy draftsmanship or, in the case of the electronic computer, by improper equipment or incompetent personnel. The neglect of some seemingly inconsequential detail can spell the difference between a good or bad chart. The term "mechanics," as used here, implies much more than simply the routine task of drawing lines with pen and ink and mounting paste-ups on drafting material. Rather, at this stage, important design decisions based on expertise and sound judgment still must be made. The mere routine of rendering a chart, whether manually or by electronic computer, after the basic design decisions have been determined represents a very different dimension.

Because of space limitations, it would be impossible to cover even the most important steps to say nothing of the multitude of minutiae involved in the mechanics of preparing charts. Accordingly, brief comments concerning only a few of the salient aspects of the process must suffice. For an extensive treatment of the subject, the reader is referred to our *Handbook of Graphic Presentation*.[35] Without attempting to set down specific rules, the following details must be considered in the mechanics of preparing charts: size and proportions of the chart with particular regard to its execution, reproduction, and the medium in which it is to appear; weights and patterns of the many different lines in the chart; type, size, and location of special symbols; size, location, and type of lettering; spacing of various components, including margins; selection and location of scales; formulation and location of scale designations, labels, legends, footnotes, and explanatory notes; wording and placement of title; determining relationships and balance of elements within the chart; and checking and rechecking for accuracy of plotting and correctness of spelling, punctuation, and other details.

WHAT IS A "GOOD" CHART?

Since the obvious purpose of improving design techniques is to enhance the quality of charts, it is pertinent at the outset to describe with some degree of clarity and specificity what may properly be considered the characteristics of a "good" chart. Although such a characterization must to a considerable extent be made within the context of value judgment, nevertheless a marked consensus will be found among knowledgeable and experienced practitioners in the field. This, of course, is not unique to graphic presentation. Analogous values and standards of what is considered "good" and "bad" and "acceptable" and "unacceptable" exist in every well-established artistic, literary, scientific, and engineering field or specialty. In all likelihood, the development of genuinely objective and quantitative procedures for evaluating the quality of charts will take at least several more decades.

The qualitative or functional evaluation of charts may be conceived in terms of a particular chart with reference to the specific circumstances and conditions that have determined or at least conditioned its conception, planning, and execution, or in broader and more general abstract terms in which idealized cri-

---

[35] Calvin F. Schmid and Stanton E. Schmid, *Handbook of Graphic Presentation*. New York: John Wiley & Sons, 1979, passim.

teria are used to categorize charts as "good" or "poor."

In judging the quality or acceptability of a particular chart, consideration should be given to the degree it fulfills or is compatible with: (1) its specific purpose and function; (2) the nature of the data that it is designed to depict; (3) the characteristics of the users; (4) physical and other constraints involved in the process of constructing the chart such as size, type, and quality of reproduction; the use of one or more colors; (5) type of equipment and material available.

A consensus of criteria that are generally applicable as a basis for categorizing charts as "good" or "poor" or "acceptable" or "unacceptable" includes (1) accuracy, (2) simplicity, (3) clarity, (4) appearance, and (5) well-designed structure. The labels used to describe these criteria have of necessity been taken from common parlance and thus lack the precision and completeness of meaning that ideally should be utilized in a discussion of this kind. The labels are merely descriptive terms that pertain to certain features of charts that are considered to be well-designed and functionally effective. In recognition of these limitations, the following brief explanations and comments should be sufficient to clarify these labels.

Accuracy. The dimensional and other physical aspects of a good chart should reflect the highest degree of accuracy possible within the practical limits imposed by expert draftsmanship or the electronic computer being used. Furthermore, a chart should not be deceptive, distorted, or misleading or in any way susceptible to wrong interpretation as a consequence of inaccuracies or careless construction. Also, care should be taken not to create optical illusions by the misplacement of lines or the failure to select appropriate cross-hatching patterns and other symbols. Not only can a chart that is manifestly inaccurate or unreliable lose its credibility, but the integrity of the entire publication of which it may be a part, may be called into question by the reader. Unfortunately, statistical graphics has become a fertile field for substantial numbers of wide-ranging dabblers and dilettantes. The unceasing crop of crude and unreliable charts in newspapers, popular magazines, books, and even scientific papers and more extended reports attest to this fact.

Simplicity. The basic design of a statistical chart should be simple and straightforward, not loaded with irrelevant, superfluous, or trivial symbols and ornamentation. An important design objective is to maximize the information communicated within the limits of simplicity and conciseness. A "good" chart should never possess the features of a complex, idiosyncratic puzzle that demands a substantial amount of time and a strained frustrating effort to decipher. Statistical charts are media of visual communication whose meaning should be perceived and interpreted smoothly, quickly, and reliably. Without minimizing the importance of innovation and originality in statistical graphics, it is a fact that oftentimes, "novel" charts and "new" systems of graphical symbols turn out to be arcane puzzles that do violence to virtually every essential quality required of an effective medium of visual communication.

In this connection, the interesting concept of "noise" borrowed from electronic communication has been applied analogically to the visual communication system of cartography. "Noise" in electronics communication implies unwanted interference with clear transmission of the signal such as incorrect voice sounds in speech, static, and distortions of appearance in television. Hence, in cartography and statistical graphics, any distracting elements in a chart that inhibit effective visual communication, such as trivial and eccentric symbols, incompatible or overpowering lettering, ornate and superfluous decorations, or bizarre hatching patterns may be referred to as "noise."[36]

An excessive amount of stylizing and artistic decoration should be avoided. The lettering on a chart, for example, should display a simple, uniform style. Simplicity and uniformity, when implemented in good taste, are conducive to the appearance of unity and balance, both essential features of a well-designed chart.

Clarity. In assessing the quality, value, and effectiveness of a chart, clarity is one of the most important characteristics. If a chart is visually ambiguous and confusing, the communication process can quickly break down, no matter how superior many elements of a chart may seem. When applied specifically to a statistical chart, what does clarity mean? It means that the chart can be easily read and understood; that there is a forceful and unmistakable focus on the message that it is trying to communicate; that there is a truthful and unambiguous representation of facts; that the message it conveys is meaningful and

---

[36] Arthur H. Robinson and Barbara Bartz Petchenik, *The Nature of Maps,* Chicago: The University of Chicago Press, 1976, pp. 23–42.

comprehensible; that in a visual sense the elements of the chart represent a harmonious combination, each occupying its proper place; and that the reader is actually aided in the interpretation of facts, thus enhancing understanding as well as saving time and effort.

Appearance. The appearance of a chart has much to do with the interest or appeal it may elicit as well as the confidence it may create. A "good" chart is one that is designed and constructed to attract and hold attention by reflecting a neat, dignified, and professional appearance. It possesses "style" in the best sense of the word. It is not slovenly, dull, or eccentric. A good chart is artistic in that it embodies harmonious composition, unity, proportion, contrast, and balance. Artistic flourishes and special ornamentation are superfluous and can seriously detract from the appearance of a chart.

In this connection, it is of some significance to note that in a recent (1980) brochure published by a well-known university announcing seminars in computer graphics, the potential registrant for the program on "Principles of Effective Map Design," is given the assurance that "you will learn, not how to make 'prettier' maps, but rather how to make more useful and efficient maps."[37] Just what this strangely paradoxical statement means is not readily evident. Does it really imply that there is something inherently different or antithetical between "pretty" (an obviously denigrating usage of the word "pretty" in this context) and "useful and efficient" maps? "Pretty," that is, attractive maps, can be just as "useful and efficient," if not more so, as "ugly," that is, unattractive maps. To imply otherwise is sheer nonsense.

Well-Designed Structure. The structure of a well-designed chart should conform to certain basic principles. One of the simpler, but important principles is the full recognition that all design elements are interdependent. Another principle, considerably more complicated in actual practice, signifies that the visual importance and distinction assigned to the various elements of a chart should be commensurate with the intellectual significance of the ideas being presented.[38] The implementation of this principle is based mainly on the principle of contrast. For example, the more significant an element is, the more prominent it should be visually. There are visual hierarchies. Accordingly, based on the criterion of relative importance, symbols or other elements of a chart may vary in size. Similarly, the basis of contrast or emphasis may be indicated by variations in color or in value; by weight, pattern, texture or length of lines; by shape or form; and by position, orientation, or direction. In recent years, cartographers have found the concept of the "figure-ground relationship," borrowed from perception psychology, to be particularly useful as a design principle. The principle describes certain qualities in an image, where part, referred to as the figure, stands out conspicuously in comparison to the remainder, referred to as the ground. This principle is similar to the principle of contrast, where the more important elements of a chart are emphasized visually and the less important elements are structured so as to recede in prominence. When viewing the structural characteristics of a chart as a total visual composition, such features as shape, balance, proportion, and unity are very significant.[39]

STANDARDS OF GRAPHIC PRESENTATION

To achieve sound, logical, and consistent chart design it is important to observe carefully the guidelines and principles represented by a body of thoroughly tested and well-established standards. In statistical graphics such a body of basic standards does exist for various kinds of time-series charts, including rectilinear coordinate line charts, surface charts, and semilogarithmic charts. These standards and values are very real and meaningful and are an integral part of the discipline; they give direction to such basic elements as design criteria, practices, and techniques.

Frequently, standards develop without conscious direction through such processes as common practice, imitation, and precedent. Also, standards may be formalized and systematized through consensus by special committees or groups created for such a purpose.

---

[37] Harvard University Laboratory for Computer Graphics Announces One-Day State-of-the-Art Briefings on Today's Most Explosive New Technologies, February–December 1980, p. 12. Interestingly, this same idea was repeated subsequently in a brochure advertising Volume 17, *How to Design an Effective Graphics Presentation,* in The Harvard Library of Computer Graphics 1981 Collection.

[38] Arthur H. Robinson, *The Look of Maps,* Madison: The University of Wisconsin Press, 1952, pp. 65–74.

[39] Arthur H. Robinson, *The Look of Maps,* Madison: The University of Wisconsin Press, 1952, pp. 65–74.

The first formalized and generally accepted standards in graphic presentation were published in 1915 by a joint committee representing more than a dozen American associations and agencies. Willard C. Brinton was the moving spirit behind this endeavor, and the sponsoring organization was the American Society of Mechanical Engineers. The 1915 report was relatively brief, consisting of 17 simply stated basic rules, each illustrated with one to three diagrams.[40]

Since the publication in 1915 of the report by the original joint committee, other committees prepared greatly expanded reports on standards of graphic presentation in 1936, 1938, 1960, and 1979. These reports have proved very valuable, but as indicated earlier, they are limited in application since they are devoted exclusively to time-series charts.[41]

Although standards for all the manifold graphic forms not classified as time-series charts have not been explicitly organized and sanctioned through collective action by a special committee or organization, they do exist and are an integral part of the discipline. They simply have not been formally organized and codified through discussion, evaluation, and consensus by a committee or group. For the most part, these standards are commonly embodied in textbooks and manuals and, in addition, are reflected in the work produced by knowledgeable and proficient specialists. The value and utility of the standards thus presented depend on the fidelity with which the standards conform to the best state of knowledge relating to the theory and practice of graphic presentation.

As standards become more explicit and are formalized through rational evaluation and consensus, graphic presentation can rid itself more easily of uncertainties and inconsistencies. Standards should never be treated as ultimates. Sound standards embody the best current usage and define knowledge at a given point in time, usually by stating what is "best" when judged according to some set of criteria. When knowledge increases or criteria change, standards must and do change. As experience in the field of graphic presentation broadens and deepens and as new problems occur, changing practices are inevitable. New standards are created, while other standards may become outmoded.[42]

---

[40] Joint Committee on Standards for Graphic Presentation, "Preliminary Report Published for the Purpose of Inviting Suggestions for the Benefit of the Committee," *Publications of the American Statistical Association,* **14** (1914–1915), 790–797. This report was also published as a separate pamphlet by the American Society of Mechanical Engineers.

[41] Committee on Standards of Graphic Presentation, *Time-Series Charts: A Manual of Design and Construction,* New York: The American Society of Mechanical Engineers, 1938; Subcommittee Y15.2M of the Committee on Preferred Practice for the Preparation of Graphs, Charts and Other Technical Illustrations, *American National Standard Time-Series Charts,* New York: American Society of Mechanical Engineers, 1979.

[42] For a detailed discussion of standards as they pertain to the design and construction of statistical charts, see Calvin F. Schmid, "The Role of Standards in Graphic Presentation," *American Statistical Association, Proceedings of the Social Statistics Section, 1976,* Washington, D.C.: American Statistical Association, 1976, pp. 74–81; reprinted in U.S. Bureau of the Census, *Graphic Presentation of Statistical Information,* Technical Paper 43, Washington: U.S. Bureau of the Census, 1978, pp. 69–78.

CHAPTER TWO

RECTILINEAR COORDINATE LINE CHARTS

Problems, Principles, and Standards

OF ALL GRAPHIC FORMS, THE MOST RIG-orous and definitive design principles and standards have been formulated for rectilinear coordinate line charts. Also, in comparison to other types of charts, the rectilinear coordinate line chart is the oldest, simplest, most familiar, and most widely used. Nevertheless, in spite of these facts, rectilinear coordinate line charts have not escaped the superficialities and incompetence of poorly informed and unskilled chartmakers.[1]

Numbered among these poorly informed and unskilled chartmakers are specialists in other fields, including the physical, biological, and social sciences. Without making a sweeping indictment, it can be said that many of these specialists are responsible for large numbers of badly designed and badly executed statistical charts, not only rectilinear coordinate line charts, but other charts as well. This fact can be readily substantiated using numerous examples appearing in professional publications. Illustrations of such instances as they relate to rectilinear coordinate line charts will be presented in the course of the present discussion.

However, as a logical first step, in order to establish a sound basis for judging the quality and effectiveness of such charts, it seems appropriate to summarize certain essential design criteria, principles, specifications, and standards pertaining to rectilinear coordinate line charts. Hence Figure 2-1 is provided as an example of a fairly typical rectilinear coordinate line chart with explanatory labels indicating its essential components.

1. Coordinate Axes. The basic structure of the rectilinear coordinate line chart is derived by plotting figures in relation to two axes formed by the intersection of two perpendicular lines. The horizontal line is called the X-axis or axis of abscissas and the vertical line, the Y-axis or axis of ordinates. The point of intersection is referred to as the "origin" or "origin of coordinates." The scales are arranged in both directions, horizontally and vertically. Measurements to the right and above the origin are positive (plus), whereas measurements to the left and below the origin are negative (minus). The areas of coordinates or plotting areas

---

[1] A well-established and authoritative discussion of standards for rectilinear coordinate line charts comprises the major portion of the following report: American National Standards Committee Y15, *Time-Series Charts,* New York: The American Society of Mechanical Engineers, 1979.

are divided by the axes into four sections known as "quadrants," which are numbered counterclockwise beginning with quadrant I which is located to the right of the Y-axis and above the X-axis. Most statistical charts utilize quadrant I almost exclusively. (See Figure 2-1).

2. Scale Divisions. In a rectilinear coordinate chart, each of the two axes are marked off in equal units beginning at the point of origin. The actual spacing of the units, whatever they happen to represent, is almost invariably different for the two axes. Scale divisions for both axes are indicated by coordinate lines and by ticks or scale points. The values represented by the various scale divisions, particularly the larger ones, are indicated by figures.

3. Coordinate Lines or Grid Lines. Coordinate or grid lines should be kept to a minimum, but there always should be a sufficient number to read clearly and accurately the values represented by the curves on the chart. Coordinate lines should be comparatively light, in contrast to the curves, which should stand out prominently from the coordinate background.

4. Zero Line or Other Line of Reference. In order to give emphasis to the zero line or other line of reference such as a 100-percent scale line, it is common practice to make such lines heavier than

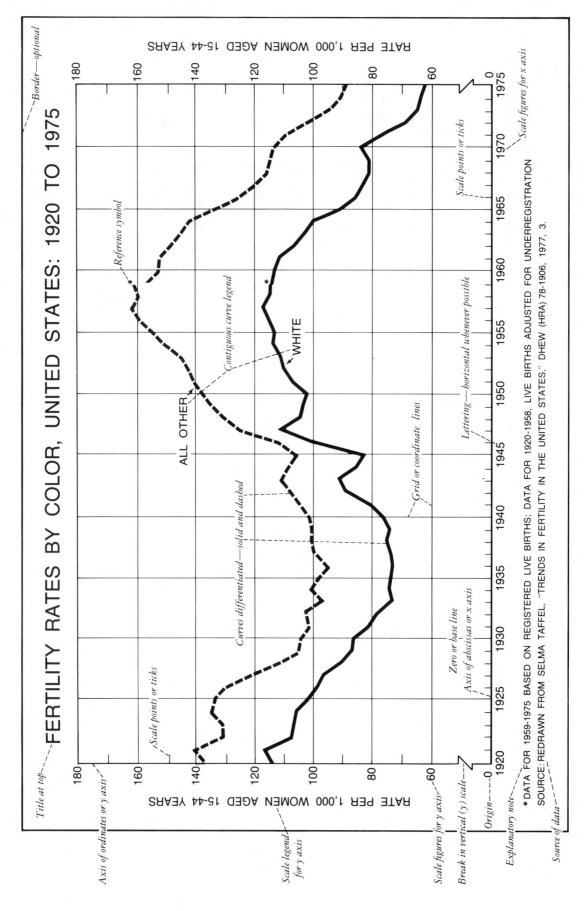

Figure 2-1. *Chart designed as a model to illustrate the essential elements or components of a rectilinear coordinate line chart.*

other coordinate or grid lines. It will be observed from Figure 2-1 that the first coordinate value on the ordinal axis is zero, and this coordinate line has been made noticeably heavier than other coordinate lines on the chart.

5. Choice of Scales and Chart Proportions. The scales chosen for both axes should result in a well-proportioned chart. That is so say, the resultant scales should not minimize or exaggerate variations in the curves. Unfortunately, this is frequently done either wittingly or unwittingly by sharply expanding or contracting the scales.

6. Break in Vertical Scale. Sometimes when only the upper portion of a coordinate field is necessary to portray the data, it is permissible to eliminate the lower portion of the field. Since it is imperative to retain the zero or base line, a break must be shown on the vertical scale at a point that will not interfere with the remaining coordinate field required for plotting the data. (See Figure 2-1).

7. Scale Figures. The scales and scale figures run from left to right on the horizontal scale and from bottom to top on the vertical scale. Scale figures are generally located fairly close to full coordinate lines, except when data are plotted to space on the horizontal axis, the scale figures are placed between full coordinate lines. In designating scale values, 5 or multiples of 5 are frequently used in preference to such units as 3, 6, 7, 9, 11, and so on. On the vertical axis when the scale values run into larger numbers such as hundreds or thousands, ciphers may be omitted, and the scale would then read 5, 10, 15, 20, and so on, with a designation such as "hundreds of dollars," or "figures in thousands."

8. Scale Designations or Scale Legends. Usually both the vertical and horizontal axes carry not only figures but also explanatory designations or legends. Scale figures and scale designations for the ordinal axis usually are placed at the left side of the grid. However, when the chart is very wide, in order to avoid confusion, they may be placed on both sides of the grid. The scale designation or legend indicates what the scales represent. Where the meaning of the scale is obvious, as in the case of the horizontal axis in Figure 2-1, the scale designation may be omitted. The scale designation for the vertical axis is never or very seldom omitted.

9. Curves and Curve Labels. If there is more than one curve on a chart, the curves should be differentiated by distinctive patterns or colors along with specific labels. In Figure 2-1, one of the two curves (white population) is shown by a full, fairly heavy black line and the other curve (all other) by a fairly heavy dashed line. It is important to recognize that superimposing too many curves on a rectilinear coordinate chart may be confusing and hence self-defeating. This is especially true if the curves tend to clump together and cross and recross one another. Curve labels placed contiguous to curves are preferable to a special legend or key located in a corner of the grid or above or below the grid. Curve labels should be simple and self-explanatory.

10. Title. Every chart should have a clearly worded and succint title. Generally, the title should answer three questions: What? Where? and When? Of course, in papers or books devoted exclusively to a particular geographic locale, it would be superfluous to repeat the name of the area in the title of every chart. Customarily, the title should be placed at the top of the chart, although it may be permissible to place it at the bottom, especially if the title is set in type by the printer.[2]

EXAMPLES OF SUBSTANDARD RECTILINEAR COORDINATE LINE CHARTS

The following eight charts (Figures 2-2 to 2-9) have been selected more or less at random to illustrate common errors and deficiencies found in rectilinear coordinate line charts. All the charts were taken from various professional publications. Many more examples, of course, could be presented, but they would be superfluous, and merely serve to belabor the clearly demonstrable (if not shocking) need for more and better education in statistical graphics. For the sake of brevity, the more obvious errors and deficiencies of each of the eight charts will be listed in simple and abbreviated form, with a minimum amount of discussion.

[2] A more extensive discussion of the various aspects of rectilinear coordinate charts, including design principles and specifications may be found in Calvin F. Schmid and Stanton E. Schmid, *Handbook of Graphic Presentation,* New York: John Wiley & Sons, 1979, pp. 31–60.

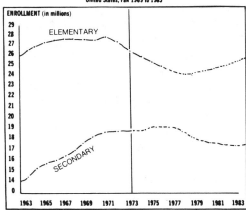

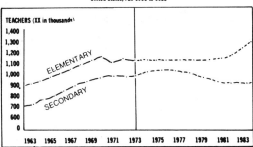

Figure 2-2. Examples of poorly planned and poorly executed rectilinear line charts. Among the more obvious deficiencies are the complete absence of grid lines and scale points. See text for additional comments (*From National Center for Education Statistics, HEW, Selected Statistical Notes on American Education, Washington, D.C.: Government Printing Office, 1975 p. 9.*).

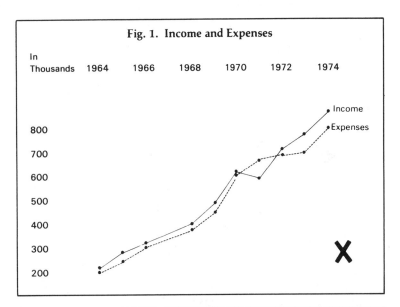

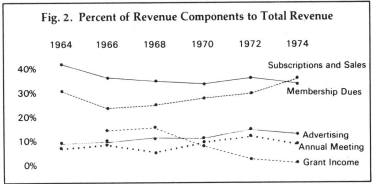

Figure 2-3. Additional examples of substandard rectilinear coordinate line charts. The most flagrant violation in this chart is the complete absence of the two axes, including coordinate lines and scale points. [*From American Sociological Association, "Financial Picture Revealed; Problem Areas Addressed," Footnotes, 3 (5) (May 1975), 8.*]

Figure 2-2 portrays trends and forecasts in elementary and secondary school enrollment, along with trends and forecasts for classroom teachers.

1. In general, the chart reflects poor planning and design with respect to the amount and size of reduction. The chart is distinctly weak and unimpressive, and most of the lettering, because of overreduction, is difficult to read.

2. From the point of view of graphic principles and standards, the most glaring shortcoming is the total lack of grid lines and scale points.

3. The curves and curve patterns indicate mediocrity in execution.

4. Although an attempt has been made to differentiate portions of the curves based on recorded statistics from those representing forecasts, the results are neither clear nor skillful.

Figure 2-3 is an attempt to show graphically trends in income and expenditures from 1964 to 1974 for the American Sociological Association.

1. Both parts of the chart are devoid of axes, except for the disparate and anomalous scale figures.

2. In the upper portion of the chart, the original axis begins with 200 and without any indication of a break in the scale. Since there are no tangible axes, "breaks" in a literal sense would not be possible. The vertical scale designation is "in thousands," that is, "in thousands of dollars." The scale figures for the abscissal axes in both upper and lower portions of the chart are for some unknown reason located at the top.

3. As in Figure 2-2, there are no grid lines or scale points, a flagrant violation of graphic principles and standards.

4. The curves and curve patterns are nondescript.

5. The overall proportion of the lower grid (Figure 2) is ill-chosen in that the height is too small in relation to its width. A ratio of at least 2 to 3, preferably greater, would be more appropriate.

Figure 2-4 has been designed to show trends and differentials in electoral participation for three racial

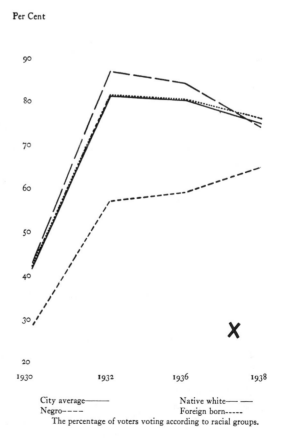

Figure 2-4. This chart also fails to meet acceptable minimal standards of chart design. Specific shortcomings are indicated in the text. (From Edward H. Litchfield, Voting Behavior in a Metropolitan Area, University of Michigan Governmental Studies No. 7, Ann Arbor: University of Michigan Press, 1941, p. 11.)

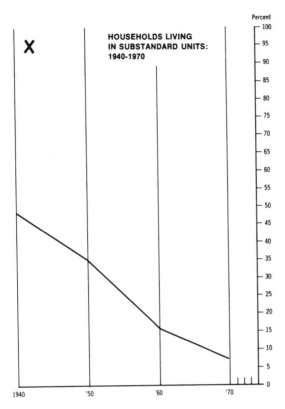

Percent
- 100
- 95
- 90
- 85
- 80
- 75
- 70
- 65
- 60
- 55
- 50
- 45
- 40
- 35
- 30
- 25
- 20
- 15
- 10
- 5
- 0

X   HOUSEHOLDS LIVING
IN SUBSTANDARD UNITS:
1940-1970

1940   '50   '60   '70

Figure 2-5. The design of this chart reflects carelessness and incompetence. Although this chart is very simple, half of the grid in its vertical dimension is superfluous. In addition, other shortcomings can be observed. See text for details. (From the Executive Office of the President, Office of Management and Budget, Social Indicators, 1973, Washington, D.C.: Government Printing Office, 1973, p. 190.)

and ethnic groups. This chart manifests some of the same deficiencies found in Figure 2-3.

1. Both the vertical and horizontal axes are non-existent except in a tenuous or imaginary sense.

2. Grid lines and scale points are nonexistent.

3. The vertical scale begins with 20 percent, not zero. There is no indication of a break in the scale.

4. The curve labels are grouped together in a "key" or "legend" located underneath the main part of the chart. As previously indicated, this arrangement is much less desirable than individual contiguous labels.

Figure 2-5 is a very simple line chart depicting trends from 1940 to 1970 in the proportion of households living in substandard dwelling units. Its deficiencies are basically the result of clumsy design.

1. Just why the vertical scale extends to 100 percent when the maximal plotted figure is less than 50 percent is difficult to explain. To be sure, the chart could have been designed as a 100-percent surface chart with hatching or shading to differen-

tiate the proportions of households living in "standard" and in "substandard" dwelling units. However, as the chart now stands, half of the grid in its vertical dimension is superfluous.

2. Again, there is no valid reason to include the blank calibrated four-year space beyond 1970, since the data terminate in 1970.

3. The chart would be more compatible with accepted standards as well as being more readable if there were at least several horizontal grid lines.

4. The following two minor criticisms might be added: First, since there is ample space, it is unnecessary to apostrophize all the years except 1940. Second, the standard practice is to place the scale figures and scale designation for the vertical axis on the left side of the grid rather than only on the right side. To state that this practice is incorrect may not be warranted, but certainly it is unconventional.

Figure 2-6 presents the number of legitimate live births and the number of deaths in the city of St. Paul from 1969 to 1979. A list of errors of omission and of commission is as follows:

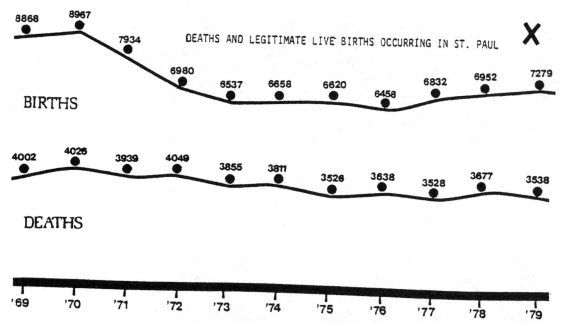

Figure 2-6.   This chart is marred by both design deficiencies and idiosyncratic characteristics. There is no vertical axis, and there are no scale figures or coordinate lines. The large black circles, misproportioned zero base line, and inappropriate lettering reflect low standards and poor craftsmanship. (From St. Paul Division of Public Health, 1979 Annual Report, St. Paul, Minnesota, 1980, p. 3.)

1. There is no vertical axis, and hence there are no scale figures and no scale legend.

2. There are no coordinate lines.

3. The zero base line is so heavy that it seems misproportioned and grotesque.

4. The large black circles seem overpowering in size, and apparently they are misplaced. If distinct black dots were to be included in the chart design, they should have been smaller in size and used as plotting points so that they would coincide with the curves. By becoming an integral part of the curves rather than being placed separately above the curves, the values represented by the dots would be more accurate. As an alternative, it would be preferable to use two distinctive curve patterns as well as different types of plotting points.

5. The lettering, especially with respect to size and style, seems careless and inappropriate.

Figure 2-7 indicates the monthly consumption of natural gas in utility generating plants in the United States for four different years, 1973 to 1976. Monthly

distributions of this kind can be plotted either to line or to space. In this chart, the data are plotted to line.

1. All the months on a chart of this kind should be indicated by coordinate lines and/or scale points. This chart is deficient in this respect.

2. The vertical axis is incorrectly calibrated. First, there is no zero referent line and no break in the scale. Second, the primary division of the vertical axis is 100, but the initial value is 160. Secondary divisions of 50, 25, or 10 would be more appropriate, and they should be indicated by grid lines and/or scale points. The last division on the vertical scale is unlabeled and oddly spaced.

3. The chart would be more easily interpreted if contiguous curve labels had been used.

4. The quality of the drafting is poor. This is reflected particularly by the four curves.

Figure 2-8 represents trends in admissions along with the resident populations of mental hospitals. A careful examination of the chart reveals three serious shortcomings in design and execution.

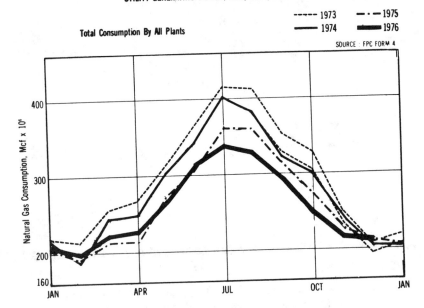

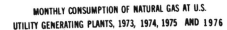

MONTHLY CONSUMPTION OF NATURAL GAS AT U.S.
UTILITY GENERATING PLANTS, 1973, 1974, 1975 AND 1976 **X**

----- 1973   — · — 1975
——— 1974   —— 1976

Total Consumption By All Plants

SOURCE : FPC FORM 4

Figure 2-7. Another poorly designed and poorly exe-
cuted rectilinear coordinate line chart. See text for a dis-
cussion of its errors and deficiencies. [From Federal
Power Commission, News Release, 10 (34-B) (week
ended August 26, 1977), p. 17.].

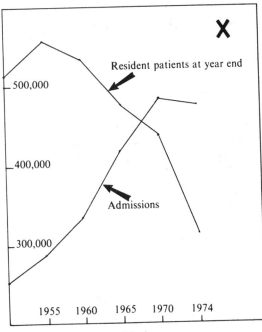

The "Revolving Door" Syndrome.

Figure 2-8. A chart that does violence to
several basic principles and standards of
rectilinear coordinate line charts. [From
Morton Kramer, "Psychiatric Services
and the Changing Institutional Scene,
1905-1985," National Institutes of Mental
Health, Series B., No. 12, Analytical and
Special Study Reports (1977), 78. Repub-
lished in University of Wisconsin, Madi-
son, Institute for Research on Poverty,
Focus, 4 (1) (Fall 1979), 13.]

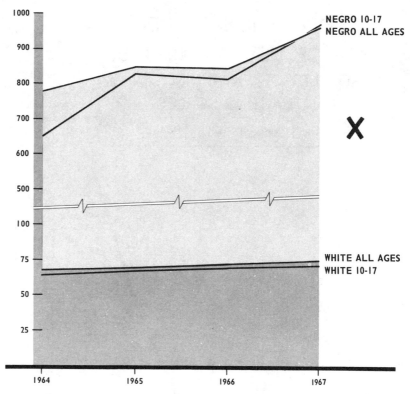

Source: UCR, 1964-67, and U.S. Census. Computations of rates done for the Task Force by the FBI. See app. 9.

—Variation in reported urban arrest rates, four major violent crimes combined, by race and age, 1964-67 [rates per 100,000 population].

Figure 2-9. Another chart replete with errors and deficiencies. See text for specific details. (From Donald J. Mulvihill and Melvin M. Tumin with Lynn A. Curtis, Crimes of Violence, Vol. 11, National Commission on the Causes and Prevention of Violence, Washington, D.C.: Government Printing Office, 1969, p. 57.)

1. No zero base line has been included on the chart. Although the first coordinate is not labeled, its value is apparently 200,000. Similarly, the last coordinate is unlabeled, and its value is apparently 600,000. There are no intermediate coordinate lines, and there are only three scale points on the ordinal axis. Strange as it may seem, the three figures for the vertical scale have been placed in the plotting field of the chart. There is no scale designation, but it can be inferred that the figures represent persons.

2. The horizontal axis, which should be as simple as possible, has been made eccentric and confusing. The time intervals represent quinquenniums, except for the last one, which is four years. Incidentally, the spacing is the same for the last four-

year interval as it is for the preceding five-year intervals. There are no vertical coordinate lines. It is not clear whether or not the two series of data begin in 1950. The right side of the chart should end exactly with 1975, not extend an indefinite space beyond the point marked 1974. In other words, the grid area should be properly framed. The scale figures for the horizontal axis have been placed in the plotting field although clearly they do not belong there.

3. Both curves are identical—thin, unimpressive lines. The heavy black arrows detract further from the chart.

Figure 2-9 attempts to show trends and differentials in urban arrest rates for whites and blacks for the total of four major violent crimes—criminal homi-

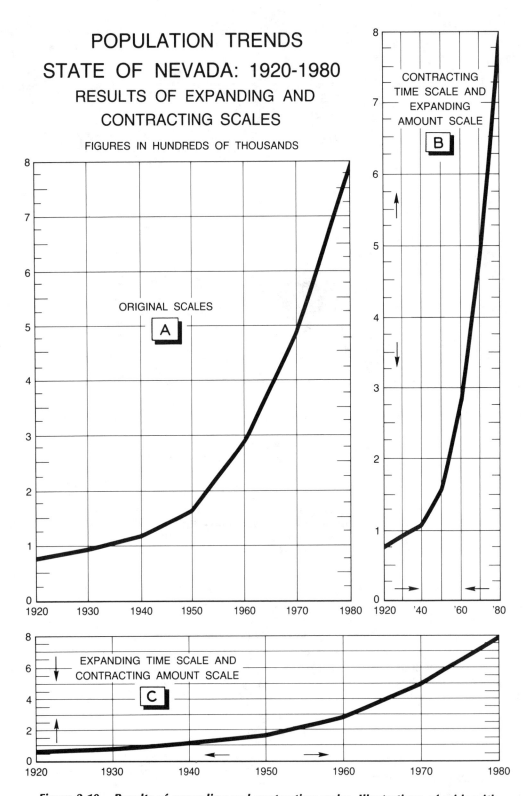

POPULATION TRENDS
STATE OF NEVADA: 1920-1980
RESULTS OF EXPANDING AND
CONTRACTING SCALES

FIGURES IN HUNDREDS OF THOUSANDS

ORIGINAL SCALES

A

CONTRACTING
TIME SCALE AND
EXPANDING
AMOUNT SCALE

B

EXPANDING TIME SCALE AND
CONTRACTING AMOUNT SCALE

C

Figure 2-10.   Results of expanding and contracting scales. Illustrations of grids with varying proportions and their influence on the configurations of curves.

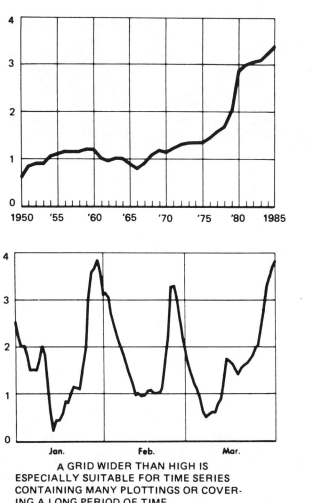

A GRID WIDER THAN HIGH IS
ESPECIALLY SUITABLE FOR TIME SERIES
CONTAINING MANY PLOTTINGS OR COVER-
ING A LONG PERIOD OF TIME.

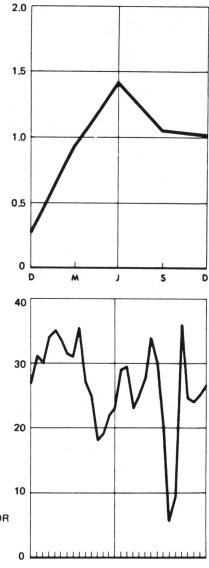

A GRID HIGHER THAN
WIDE IS SOMETIMES BETTER FOR
SERIES COVERING A SHORT
PERIOD, OR EXHIBITING RAPID
CHANGE.

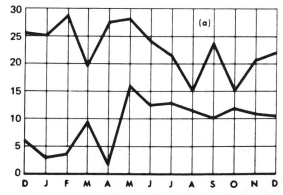

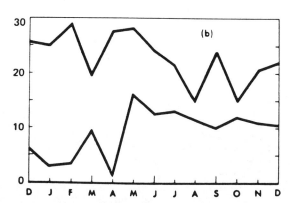

MOST TIME-SERIES CHARTS REQUIRE A TREATMENT SOMEWHERE BETWEEN THESE EXTREMES:
(a) TOO MANY GRID RULINGS. (b) NOT ENOUGH RULINGS.

Figure 2-11. This chart provides constructive suggestions for resolving two basic problems concerning rectilinear coordinate line charts: (1) grid proportions and (2) grid rulings. (From American National Standards Committee Y15, Time-Series Charts, New York: The American Society of Mechanical Engineers, 1979, p. 11.)

cide, forcible rape, robbery, and aggravated assault—from 1964 to 1967. This chart is grossly misleading because of its aberrant design and structure.

The most flagrant errors are in the vertical scale. It will be seen that the original scale begins with divisions of 25 from 0 to 100. These four units cover the entire range of rates for the white population (66.7 to 77.0). The rates for blacks ranged from 780.3 to 959.6. Strangely, there is a break in the scale and the values of the scale divisions are transformed from 25 to 100. However, the spacing of the scale divisions remains the same. To say the least, these changes are confusing, misleading, and completely contrary to design practice. Because of their serious nature, these errors alone would cause this chart to be rejected.[3]

ARE THERE OPTIMAL GRID PROPORTIONS?

In rectilinear coordinate charts, grid proportions are of pronounced significance as the determinants of the visual impression conveyed. By altering grid proportions, it is possible to show the same data in such a way that the visual effect would be entirely different. For example, if the vertical dimension of the grid is greatly elongated, the configuration of the curves would be characterized by sharp and steep trends and fluctuations. On the other hand, if the horizontal dimension is made extremely wide, the corresponding curves would tend to become flat; the fluctuations, dwarfed and relatively shallow. Alterations of this

kind indicate either the expansion or contraction of the scales, that is, whether or not their spacing has been widened or narrowed. Examples of alterations of this kind are shown in Figure 2-10. In order to emphasize the implications of such changes, the resultant charts may seem bizarre. It also indicates that a chart may be "correct" in other respects, but badly proportioned. This fact applies not only to rectilinear coordinate charts, but to other types of charts as well. Perhaps it should be pointed out that in order to emphasize or exaggerate certain movements or trends, the designer of a chart may consciously attempt to mislead the reader by distorting either the time or amount proportions of a chart. Obviously, manipulations of this kind are motivated by propagandist or other ulterior purposes.

There are no single optimal proportions for rectilinear coordinate charts, nor are there any precise, hard-and-fast rules for determining the most appropriate proportions for any given chart. Trial-and-error experimentation by means of preliminary sketches may be found helpful. The grid should be made to fit the data, not vice versa. Space and other

---

[3] It is significant to observe that Figure 2-9 is merely one of many substandard charts in the following report: National Commission on the Causes of Crime, Donald J. Mulvihill and Melvin M. Tumin (co-directors) and Lynn A. Curtis (assistant director), *Crimes of Violence*, Vol. 11, Washington, D.C.: Government Printing Office, 1969. See, for example, the charts on the following pages: 28, 55, 76, 77, 79, 87, 90, 91, 92, 93, 100, and 105.

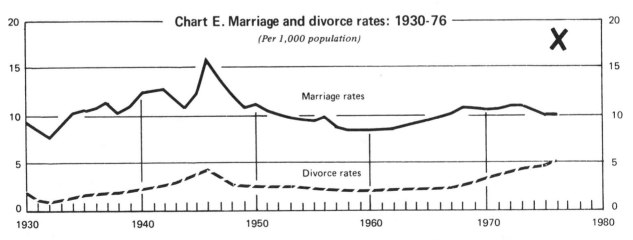

Figure 2-12.   An example of a poorly proportioned rectilinear coordinate line chart. Poor proportions may ruin an otherwise acceptable chart. (From Department of Health, Education, and Welfare, Public Health Service, National Center for Health Statistics, Facts of Life and Death, DHEW Pub. No. PHS 79-1222, Washington, D.C.: Government Printing Office, 1978, p. 10.)

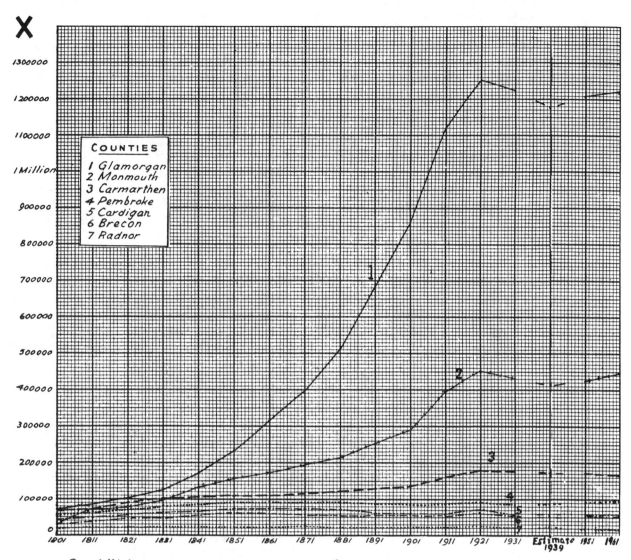

X

South Wales: Population (by Counties) 1801-1961.

Figure 2-13. A crudely designed and executed rectilinear coordinate line chart drawn on commercially printed graph paper. As it has been prepared, the chart is virtually impossible to interpret. [From William Rees, An Historical Atlas of Wales, 3rd ed., London: Faber and Faber, Ltd., 1967, Plate 70(a).]

constraints of the publication in which a chart is to appear may create additional problems.

The following suggestions taken from the 1979 American National Standards *Time-Series Charts* may be found helpful.[4]

1. Oblong-shaped grids are preferable to square grids. Good standard proportions are 2 to 3, and 3 to 4, but almost any proportion is satisfactory if it helps to present the data effectively.

2. A grid wider than it is high is likely to be bad for most time series. However, sometimes a grid higher than it is wide may be preferable for (a) a relatively short series or one in which there are few plotting points; (b) series with a strong trend or rapid change; (c) overlapping curves that would "run together" otherwise; and (d) important fluctuations that would not be noticeable on a horizontally arranged grid. (See Figure 2-11.)

---

[4] American National Standards Committee Y15, *Time-Series Charts*, New York: The American Society of Mechanical Engineers, 1979, p. 10.

Figure 2-12 is an example of a poorly proportioned chart taken from a recently published report. The ratio of height to width creates an imbalance, and as a consequence, the curves are relatively flat. Even the peaks and valleys in the marriage rates during the 1930s and 1940s do not stand out very clearly. These fluctuations as well as the few shown by the divorce curve would be much more prominent and meaningful if the vertical dimension had been made considerably larger. A ratio of height to width of say, about 3 to 4, or about 2 to 3, rather than its present ratio of slightly less than 1 to 3 would have produced a more realistic and significant chart.

In this connection a statement made by Willard C. Brinton in the early 1900s still holds true today.

The beginner in curve plotting and in curve reading is apt to be somewhat puzzled by the different effects which may be obtained by changing the ratio between the vertical scale and the horizontal scale. . . . Just as the written or spoken English language may be used to make gross exaggerations, so charts and especially curves may convey exaggerations unless the person preparing the charts uses as much care as he would ordinarily use to avoid exaggeration if presenting his material by written or spoken words. Most authors would greatly resent it if they were told that their writings contained great exaggerations, yet many of these same authors permit their work to be illustrated with charts which are so arranged as to cause an erroneous interpretation. If authors and editors will inspect their charts as carefully as they revise their written matter, we shall have, in a very short time, a standard of reliability in charts and illustrations just as high as now found in the average printed page.[5]

IS IT POSSIBLE TO DETERMINE THE OPTIMAL NUMBER OF GRID LINES?

As a general rule, the number of coordinate or grid lines should be kept to a minimum. For certain mathematical or engineering charts, it may be expedient to space the grid lines close together, but generally for graphic presentation, a relatively large number of

grid lines would be superfluous. Moreover, they could result in clutter and detract from the visual impact of the chart. In this connection, it is inadvisable to use printed graph paper in the preparation of charts except perhaps for preliminary sketching or layouts. Printed graph paper with its fixed dimensions and myriad of irrelevant and useless lines lacks the flexibility essential to the preparation of well-designed charts. (See Figure 2-13).

Grid lines serve an important purpose by guiding the eye in locating and reading values on curves and on other points in a chart. In this respect a good rule to follow is that, "There should be as many horizontal rulings as needed to show the reader the amount value of the plotted points *to the degree of accuracy required.*"[6] As a supplementary rule, there should be a coordinate or grid line for each labeled division on the ordinal axis. Also, there should be an adequate number of grid lines to differentiate the major divisions on the abscissal axis. Figure 2-11 shows sketches of two charts categorized as "extreme" examples: One has too many grid lines, and one has no grid lines. In aiming for an optimal balance of grid lines, the number of grid lines should be somewhere between these extremes. The final decision, of course, must be left to the chartmaker. The size of the chart, the type and range of the data, the number of curves, the length and detail of the period covered, as well as other factors will exert some influence on any decision that is made.

SPACING AND OTHER PROBLEMS IN DESIGNING THE HORIZONTAL SCALE: THE DANGERS OF IRREGULARITIES AND AMBIGUITIES

In a rectilinear coordinate line chart, the horizontal axis always represents time, the so-called independent variable. However, because of ill-defined, shifting, and ambiguous time units along with gaps and errors in statistical data, the chartmaker may experience difficulties in designing and adjusting scale intervals and in plotting data correctly. For example, in the compilation of statistical data, a "year" could mean a calendar year or a fiscal year. Moreover, a fiscal year may be defined differently by different orga-

---

[5] Willard C. Brinton, *Graphic Methods for Presenting Facts*, New York: The Engineering Magazine Company, 1914, pp. 352–353.

[6] American National Standards Committee Y15, *Time-Series Charts*, New York: The American Society of Mechanical Engineers, 1979, p. 10.

SUGGESTIONS FOR TIME-SCALE RULINGS AND DESIGNATIONS

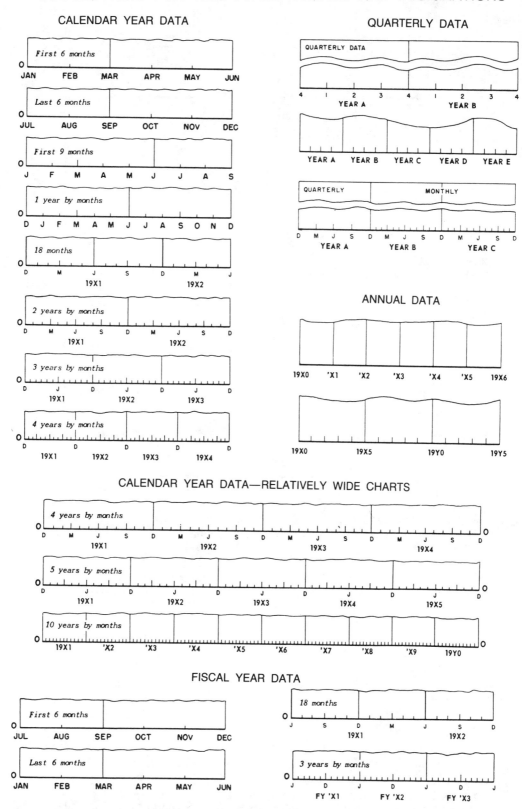

Figure 2-14.   *It will be observed from these sketches that in general, the size of the chart, length of the period covered, and nature of the time interval determine the number of rulings and scale designations for the time axis of rectilinear coordinate charts. For example, every month should be labeled for periods up to 1 year; every third month for periods from 15 to 24 months. Relatively wide charts should have more scale labels. (From Department of the Army, Standards of Statistical Presentation, Department of the Army Pamphlet 325-10, 1966, pp. 86–87.)*

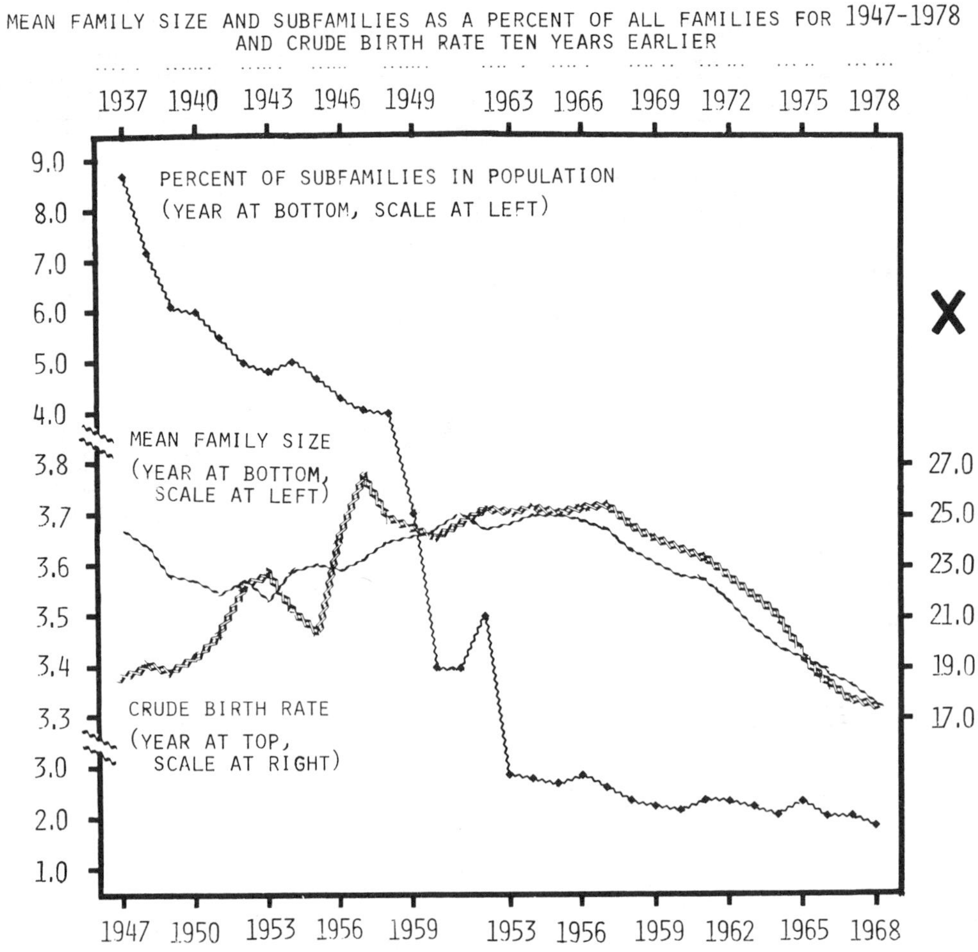

MEAN FAMILY SIZE AND SUBFAMILIES AS A PERCENT OF ALL FAMILIES FOR 1947-1978
AND CRUDE BIRTH RATE TEN YEARS EARLIER

Figure 2-15.   A multiple-scale rectilinear coordinate line chart with confusing and anomalous deficiencies. [From Judith Treas, "Postwar Trends in Family Size," Demography, 18 (August 1981), 321–334.]

nizations. A "day" may mean a workday, a store-day, or a 24-hour day. Statistical data are sometimes tabulated for specified periods of time (period data) and sometimes for specified points in time (point data). Again, the basic time unit in a series of data may be shifted from years to quarters, or from quarters to months, and so on. Figure 2-13 illustrates a number of scale and plotting problems as a consequence of irregular time intervals, period and point data, omissions of data, and shifting time intervals.

Figure 2-14 consists of several typical horizontal scale designs, mainly for plotting data on the line. These sketches illustrate how horizontal scales with different kinds of time intervals of varying number might be laid out. In most examples, the basic time unit is the month, and the total length of time covered ranges from six months to ten years. Figure 2-14 also illustrates how months as a basic time unit may be

grouped by calendar years as well as by fiscal years. In addition, some of the sketches include quarterly and yearly time intervals.

Although, as a general rule, it is simple to plot data on the line, sometimes it may be necessary or even preferable to plot data by space. In such instances the usual plotting point is at midspace of the basic time interval such as say, week, month, quarter, or year. When plotting by space, labels on the horizontal scale are placed between grid lines or ticks, rather than being centered on grid lines or ticks, as in line plotting.

---

[7] Kenneth W. Haemer, "Double Scales Are Dangerous," *The American Statistician*, **II,** 3 (1948), 24. Also, for certain other details see American National Standards Committee Y15, *Time-Series Charts,* New York: The American Society of Mechanical engineers, 1979, pp. 21–22.

MULTIPLE SCALES CAN BE DANGEROUS

In designing rectilinear coordinate line charts with multiple scales, it is important to be aware of the problems and pitfalls that may be involved. In the first place, unless the chart is constructed properly, the resultant message may be distorted and confusing and in the second place, even though a chart of this kind may be technically correct, it is easy to misread and misinterpret. The average reader is not sufficiently familiar with statistical charts to understand charts with multiple scales. Charts with two scales may be found difficult to interpret; charts with three or more scales may be utterly impracticable in this respect.

Apart from the risks involved in using multiple scales, they seem to offer logical and sometimes useful solutions to the problems of: (1) bringing together and "opening up" for comparison curves that vary widely in magnitude and otherwise would be far apart on a single-scale chart; (2) comparing two or more variables measured in different units that cannot be compared conveniently on a single-amount scale; and (3) comparing two or more series without computation. The last problem can be solved by using the point of juncture of the two curves as the base point. This has the added advantage that, since absolute-value scales are used instead of the percentage-value scales, ready reference can be made to actual magnitudes.[7]

Figure 2-15 is an example of a poorly designed and poorly executed rectilinear coordinate line chart with multiple scales. It attempts to portray trends and relationships among three variables—"mean family size," "crude birth rate," and "percent of subfamilies in population." However, the chart contains serious errors and its interpretation is distracting and confusing. First, both the time scales—the one at the top and the other at the bottom—lack continuity. Apparently, the last six designations (years) of both scales have been misplaced and should be interchanged. Is this discrepancy a mechanical (drafting) error or is it a basic design error? Second, the vertical scale with uniform unit spacing exhibits two breaks, each of which is followed by a shift in scale values. Before the first break, the scale runs from 1.0 to 3.0. Also, it will be noted that there is no origin. Following the break, the scale values are changed to tenths, but the spacing of the scale units remains the same. The values for this part of the scale run from 3.3 to 3.8. After the second break, the values revert to whole numbers ranging from 4.0 to 9.0. This entire procedure is erroneous and misleading. Third, the plotting of the curve showing "percent of subfamilies in the population" is incorrect. The curve begins at the upper portion of the chart and runs through two scale breaks and through changes in scale values. It would be impossible to make any kind of plotting adjustments in view of these clumsy and arbitrary shifts and breaks in the vertical axis. Fourth, although the two curves—"mean family size" and "crude birth rate"—seem to be neatly juxtaposed, this was accomplished arbitrarily, largely at the expense of an aberrantly designed chart. In addition to the shortcomings in basic design, the drafting quality and minor design features of the chart are at best mediocre.[8]

TOO MANY CURVES OR OTHER SYMBOLS

A chart overloaded with curves or other symbols can turn out to be self-defeating and worthless, or even worse than worthless, as a medium of visual communication. Because of its complexity and clutter such a chart may be extremely difficult to interpret, and instead of elucidating a problem, it may become a time-consuming and frustrating liability. There are no easy or precise recommendations to avoid this pitfall. It is not possible to prescribe definitive rules or criteria that might serve as guides for determining the optimal number of curves or other symbols for any particular graphic form, since so many factors are involved in the design process. As has been pointed out in previous discussions, in planning a chart the designer must be guided by such general criteria as interpretability, clarity, reliability, and simplicity.

Figure 2-16 was designed to portray trends in sex ratios (number of males per 100 females) of the population of the United States from 1940 to 1970. The population was divided into two broad racial categories (white and nonwhite) and each racial category was further divided into 12 age groups, making a total of 24. Each of the 24 age groups is represented by a curve in Figure 2-16, with separate grids for the white and nonwhite categories. With the resultant overlapping, intertwining, and partial concealment of several of the curves, the chart is difficult to interpret. As it now stands, it is too complex and confusing to be rated as of acceptable quality on the basis of clarity, simplicity, reliability, and interpretability.

---

[8] For another example of how a multiple-scale chart can be misused and misinterpreted, see Chapter 3. This example involves columns rather than curves, but the basic graphic system represents rectangular coordinates.

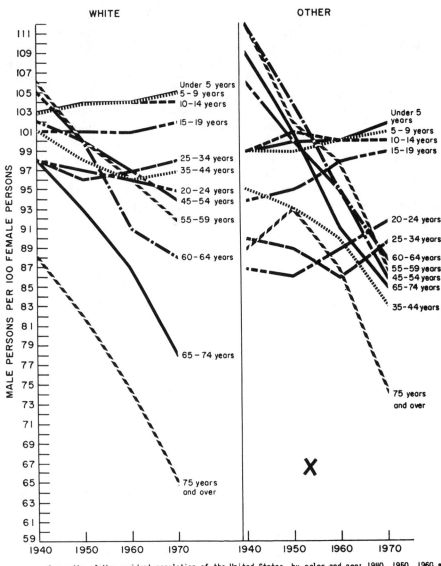

WHITE   OTHER

Sex ratio of the resident population of the United States, by color and age: 1940, 1950, 1960, and 1970.

Figure 2-16.   A rectilinear coordinate line chart with a total of 24 curves. There are 12 curves on each of the grids. The inevitable crisscrossing and overlapping of so many curves plotted within a relatively narrow range of values make this chart difficult to interpret. (From A. Joan Klebba, Leading Components of Upturn in Mortality for Men, United States: 1952–1967, *DHEW Publication No. HSM 72-1008, Rockville, Md.: National Center for Health Statistics, 1971, p. 3.)*

Obviously, the major deficiency of the basic design is too many curves in too small a space. One possible solution would be to regroup the race and age categories on four instead of two grids. Incidentally, greater emphasis should be placed on 100 as a base line, since the sex ratios are expressed as either above or below 100 (number of males per 100 females). Also, the primary calibrations and scale figures on the ordinal axes should be designated as even rather than odd numbers.

QUALITATIVE SIGNIFICANCE OF THE MECHANICS OF CHART CONSTRUCTION

As indicated in Chapter 1, the charting process involves two basic and closely interrelated dimensions: The first dimension represents the application of principles and practice of chart design; the second, the techniques and mechanics of chart construction. Not infrequently, in actual practice it is difficult if not impossible to draw a sharp line of distinction between these two dimensions.

This brief digression is designed to heighten the reader's awareness of the qualitative significance of the mechanics of chart construction in the charting process. Although chart design is of primary importance, the role of the mechanics of chart construction should in no way be minimized. Two examples of rectilinear coordinate line charts are included in this discussion to illustrate how, apart from chart design,

the mechanical dimension can have a crucial impact on the quality of statistical charts. Of course, examples analogous to the two rectilinear coordinate line charts could be multiplied manifold for every graphic form included in this book, but it would be repetitious as well as superfluous to include additional illustrations.

The essential problem of chart design is to select carefully and organize effectively a composite of interrelated elements into an optimal conceptual and

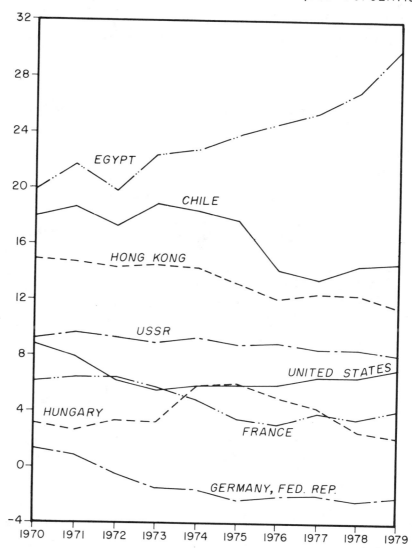

EXCESS OF BIRTHS OVER DEATHS PER 1,000 POPULATION

Figure 2-17. Although the basic design of this chart may be considered of acceptable quality, it lacks the appearance, forcefulness, and professionalism of a superior chart. These deficiencies are attributable largely to mediocre draftsmanship. Relatively minor deficiencies in the mechanics of construction can severely depreciate the quality of a statistical chart. [From Office of Population Research, Princeton University and Population Association of America, Population Index, 47 (3) (Fall 1981), outside back over.]

SOURCES: Data were taken from the following publications of the U.N. Statistical Office: Statistical Papers, Series A, Vol. 33, No. 2 (April 1981); Demographic Yearbook 1979, Tables 9 and 18; and Demographic Yearbook 1978: Historical Supplement, Table I.

physical structure for the purpose of conveying through the process of visualization a specified message as well as fulfilling other predetermined objectives; for example, these predetermined objectives may include appropriate adaptations and adjustments to the conditions under which the chart is to be used as well as to the interests and abilities of the users. Thus the task of the designer is to produce a thoroughly planned layout, which frequently must be supplemented with verbal directions and specifications. The next stage in the charting process, which is primarily mechanical, occurs when the basic design layout is transformed into final finished form, either manually or by electronic computer. Usually more than one person is involved in these two steps. Clearly the production of superior charts demands the highest standards in both chart design and chart construction. The most impeccably designed chart

Figure 2-18.   Another example of a rectilinear coordinate line chart with a basic design of average quality. Its shortcomings are due mainly to poor execution. Note especially the small-sized lettering, the weak and unimpressive curves, the location of the curve legend, and the complete omission of grid lines. A chart that is substandard primarily because of poor draftsmanship. (From First Marriages: United States, 1968–1976, DHEW Publication No. PH5 79–1913, Hyattsville, Md.: National Center for Health Statistics, 1979, p. 15.)

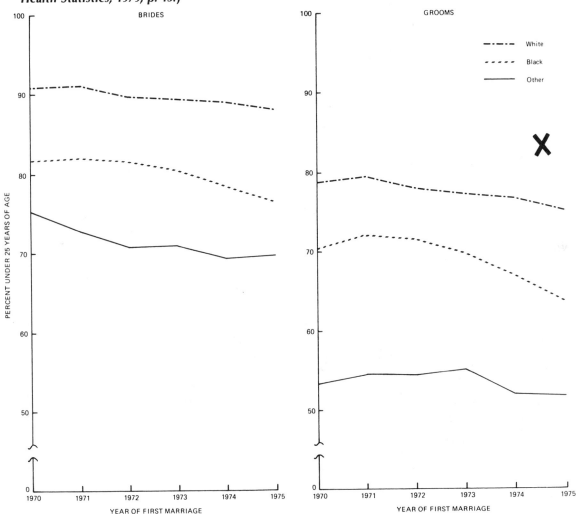

Percent of brides and grooms under 25 years of age at first marriage, by race: Reporting States, 1970-75

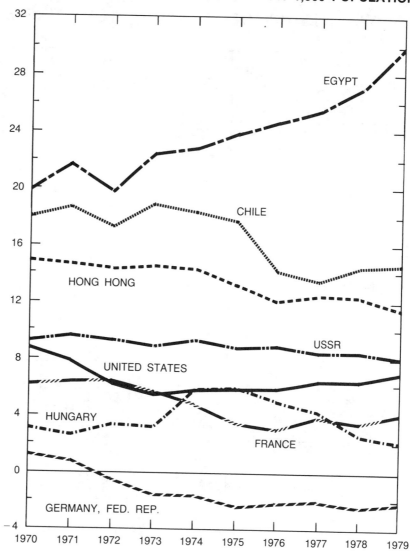

EXCESS OF BIRTHS OVER DEATHS PER 1,000 POPULATION

Figure 2-19. This chart is a redrafted version of Figure 2-17. An attempt has been made to improve the mechanical shortcomings exhibited by Figure 2-17.

SOURCES: Data were taken from the following publications of the U.N. Statistical Office: Statistical Papers, Series A, Vol. 33, No. 2 (April 1981); Demographic Yearbook 1979, Tables 9 and 18; and Demographic Yearbook 1978: Historical Supplement, Table 1.

can be ruined by sloppy draftsmanship or, in the case of the electronic computer, by improper equipment or incompetent personnel. On the other hand, the inherent inferiority of a poorly designed chart cannot be masked or dissimulated merely by superb drafting skill or extraordinary computer expertise.

Sometimes the difference between a really "good" chart and an "average" chart, or even between a "good" chart and one that is "mediocre" or "poor," may be the result of seemingly slight deficiencies in design or in construction, such as too many or too few

grid lines, the omission of an important referent line, lettering that is too small or too large, inappropriate cross-hatching or shading, lines that are too heavy or too light, or misplaced legends and labels.

Figures 2-17 and 2-18 are examples of charts with reasonably acceptable basic designs, but with mechanical deficiencies. Moreover, it can be said that they lack a "professional touch." It will be observed from Figures 2-19 and 2-20 that the charts have been redrafted for the purpose of illustrating how, in many cases, the quality of charts can be improved by rela-

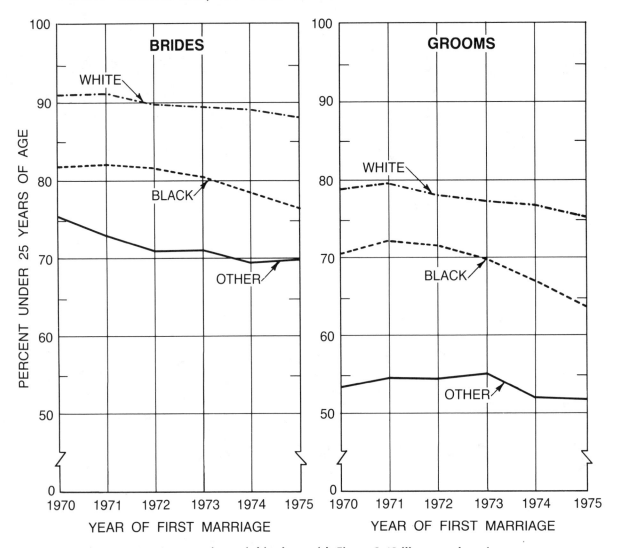

BRIDES AND GROOMS UNDER 25 YEARS OF AGE AT FIRST MARRIAGE, BY RACE: REPORTING STATES, 1970-75

Figure 2-20.   A comparison of this chart with Figure 2-18 illustrates how important the mechanical features are in considering the overall quality of a statistical chart.

tively minor mechanical (as distinguished from basic design defects) revisions and additions. In Figure 2-19, the alterations include: (1) changes in the style, size, and location of lettering,; (2) addition of zero base line; (3) relocation and addition of scale points; (4) revision of curve weights and patterns; and (5) revision of line weights for both axes. Similarly, the

more significant revisions made in Figure 2-20 are: (1) addition of lines for both horizontal and vertical axes (top and right), including scale points, (2) enlargement of all lettering, (3) shifting title to top of chart, (4) moving curve labels, (5) revision of curve weights and patterns, and (6) increasing weight of horizontal axes.

THIS CHAPTER IS DEVOTED TO SPECIAL issues and problems pertaining to two closely related families of charts: bar and column charts. Frequently, in the lay mind, probably because of their physical resemblance, no distinction is made between bar and column charts. Accordingly, the label "bar chart" is frequently applied to both.

However, it is the contention of this book that a practical and meaningful distinction between the bar chart and the column chart does exist, based largely on the characteristic arrangement of "bars" and "columns," and especially on their application. Generally, in a bar chart the "bars" are arranged horizontally; in a column chart, the "columns" are arranged vertically. More importantly, with respect to application and emphasis, the bar chart characteristically is used for direct comparisons of magnitude for descriptively labeled categories, while the column chart is used for comparisons of one or more series of variables over time. This implies that the bar chart has only one scale, but the column chart, like the rectilinear coordinate line chart, has two scales. The column chart, in comparison to the line chart, can provide greater emphasis for relatively short periods of time in portraying amounts in a single time series and greater contrast in portraying amounts in two or three series.

This distinction between bar and column charts is not an absolute, invariant rule, but rather, in most circumstances, a logical workable principle. Sometimes a reverse of this principle may occur, when columns instead of bars, or vice versa, are more appropriate or effective because of the nature of the data or because certain purposes or objectives can be more readily achieved.

The all important purpose of both bar and column charts is to portray comparisons of amounts by means of the simplest, most graphic, and most visually reliable geometric form—one-dimensional or linear bars and columns. As will be discussed later, comparisons also can be made with two-dimensional or areal forms and with three-dimensional or cubic forms.

An indispensable prerequisite for a meaningful discussion of bar and column charts is a clear understanding of their numerous variations and combinations.

TYPES OF BAR CHARTS

Verbal descriptions of eight different types of bar charts are presented in the following paragraphs. For graphic examples, see Figure 3-1.

CHAPTER THREE

BAR AND COLUMN CHARTS

Issues and Problems

Simple Bar Chart. This is one of the most useful and most widely used forms of graphic presentation. The simple bar chart is used to compare two or more coordinate items. Comparison is based on direct linear values; the length of the bars is determined by the value or amount of each category. The bars are usually arranged according to relative magnitude of items.

Bar and Symbol Chart. This is merely a simple bar chart with supplementary information indicated by a cross-line, circle, diamond, or some other symbol.

Subdivided Bar Chart. This type of bar chart, like the 100 percent bar chart described in the following paragraph, is also referred to as a "segmented" or "component" bar chart. The scale values of the subdivided-bar chart are shown in absolute numbers. To portray percentage distribution of components, the 100% bar chart should be used.

Subdivided 100-percent Bar Chart. This type of chart consists of one or more segmented bars, where each bar totals 100 percent. The various divisions of the bars represent percentages of the whole.

Grouped Bar Chart. This type of chart is also referred to as a "multiple-unit" bar chart. Comparison of items composed of two, and sometimes three units or categories can be made by this type of chart.

Paired Bar Chart. This chart, along with the deviation bar chart and the sliding bar chart (next items), is a special type of "bilateral," "two-way," or "two-directional" chart. Different units and scales can be used for each set of bars.

Deviation Bar Chart. Note that bars extend either to the left or to the right of the same base line.

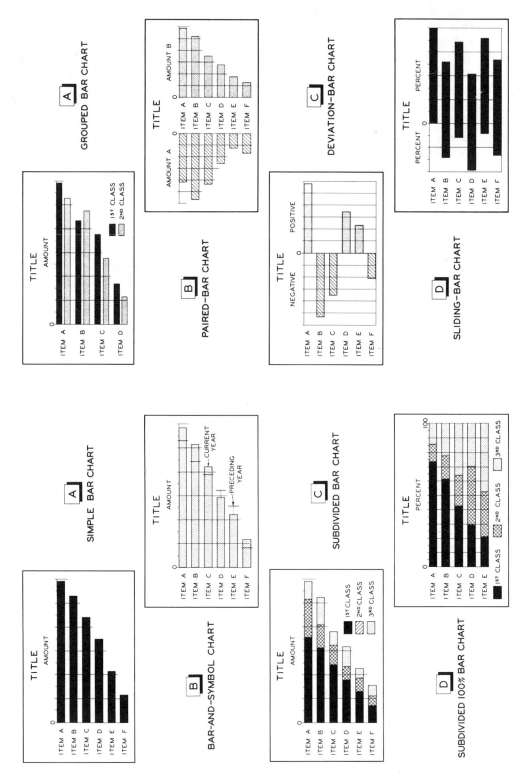

Figure 3-1. Eight different types of bar charts. See text for verbal description. (From Calvin F. Schmid and Stanton E. Schmid, Handbook of Graphic Presentation, New York: John Wiley & Sons, 1979, pp. 62–63.)

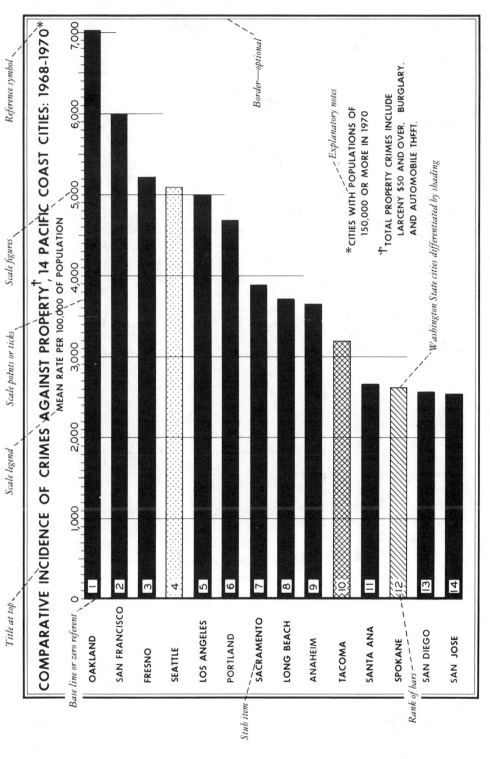

Figure 3-2. *Basic structural characteristics and design specifications of a simple bar chart. A detailed as well as a more general discussion of design problems pertaining to bar charts will be found in the text.*

This type of bar chart is especially valuable for presentation of positive-negative and profit-loss data.

Sliding Bar Chart. This is a bilateral chart in which each bar represents the total of two main components. One part of the bar is to the left of and the other part is to the right of a base line. The scale may represent either percentages or absolute numbers.[1]

BASIC STRUCTURAL CHARACTERISTICS AND DESIGN SPECIFICATIONS FOR BAR CHARTS

The basic structural elements of a simple bar chart are shown in Figure 3-2. The following explanatory statements will help to clarify the significance of each of these elements, including their relationships to one another.

1. Bars. It is only the length of the bars that possesses statistical significance since the length of each bar is determined by the value it represents. The width of the bars and intervening spaces do not have any special meaning, except insofar as they might influence the harmonious relationship and balance of the various elements of the chart, as well as the overall size and proportions of the chart. In designing a bar chart, the bar should not be excessively wide or narrow or disproportionately long or short. Decisions of this kind are determined largely by judgment and experience. There are few hard-and-fast rules to follow. As a working principle, the interspaces can be from one-fifth to one-half the width of the bars. Except for rare exceptions, the widths of both the bars and the interspaces are uniform and constant. Normally, only in area bar charts do the widths of bars vary.

In multiple bar charts and other special types of bar charts, the problem of planning optimal bar widths and spacing may present problems that are different from and perhaps more complicated than those that occur in the single bar chart. For example, in grouped bar charts, bars may touch laterally or actually overlap.

2. Ordering of Bar Charts. To facilitate comparison and analysis, it is essential that bars be arranged by some systematic order. The particular order chosen should be determined by the purpose at hand. The most common criterion in ranking bars is size or magnitude, although alphabetic, temporal, or geographic criteria sometimes may be found appropriate. Occasionally, subjective criteria such as "best" to "worst" or vice versa may be used as a basis for ranking bars. As an arbitrary rule, it is common practice to place categories such as "unclassified, "miscellaneous," and "all other" at the bottom of a series of bars.

3. Scale. Unlike the rectilinear coordinate line chart or even the typical column chart that is used to portray time series, the bar chart has only one scale. The purpose of this scale is to gauge the length of bars with reasonable accuracy. It has been suggested that a scale on a bar chart is unnecessary, since adding the original data to each of the bars would serve the same purpose. There seems to be little practical or empirical evidence to substantiate such a suggestion. As will be observed from Figure 3-2, the scale is located directly under the title of the chart. It is of the utmost importance that there is a zero line or other referent line on the scale of every bar chart. In a simple bar chart, for example, bars are always measured continuously from the zero line without any general break. This fact reflects the intrinsic one-dimensional character of both bar and column charts. Occasionally, there are instances in both bar and column charts when a "freak" bar or column is so disproportionately long that all the others are dwarfed by comparison. In such cases, it is permissible to "break" the bar or column. In so doing, the break should be neat and simple, it should be located beyond the next longest bar or column, and its value should be indicated above the bar or column. The figures on the scale should be expressed in round numbers in such units as 5s, 10s, 25s, 50s and 100s. The scale legend is placed above the scale figures. It should be short, and specific.

In designing a scale for a bar chart, care should be taken not to create an imbalance in the chart. This usually occurs when too little space is allocated to the bars. In some cases, a problem of this kind may be difficult to avoid because of awkward data, but it also may result from poor planning. In relation to the width of the body of a bar chart, at least two-thirds of the space should be allocated to the bars. Sometimes the stub labels are so long that if placed on a single line the bars would be

---

[1] Calvin F. Schmid and Stanton E. Schmid, *Handbook of Graphic Presentation,* New York: John Wiley & Sons, 1979, pp. 61–62.

crowded into such a small space that a serious im-balance would be created. Adjustments for such a contingency can be made by placing the stubs on two lines. Another type of adjustment for prob-lems resulting from awkward data is to include special inserts in the chart.[2]

Even when there is a scale, some graphic spe-cialists believe that it enhances the effectiveness of a bar chart to include a substantial amount of the plotting data on the chart. Sometimes this practice can be overdone and create clutter or optical illu-sions. Moreover, it is inadvisable to attempt to de-sign a chart for the double purpose of serving as a visual communication device and as a statistical table. However, when certain supplementary data are added to a bar chart, the preferred practice is to place them at the left end of the bars next to the zero referent line rather than inside the bars or at the right end of the bars.

4.   Shading and Cross-Hatching.   In simple bar charts, it is accepted practice to make the bars solid black or to use suitable shading or cross-hatching. When feasible, colors may be used. Shading and cross-hatching patterns should be chosen with care since some patterns are coarse, wavy, and even bi-zarre. Some patterns can create optical illusions, the most common resulting in a "tilting" of the bars or columns. Inappropriate shading or hatch-ing can seriously diminish the quality as well as the appeal of an otherwise acceptable chart. In component bar charts, where several different cat-egories may be involved, shading and hatching are essential. The simplest procedure in shading and hatching bars is to superimpose preprinted, self-adhesive tapes that are available commercially.

5.   Stubs.   The stubs or labels for the bars should be as brief as possible, but readily understood. They should conform to a straight left-hand mar-gin and be centered opposite each of the bars.

6.   Title.   Every chart should have a concise, rel-evant, and clearly worded title that answers the questions What? Where? and When? Of course, where all of the charts in a publication pertain ex-clusively to one geographic locale, it could be un-necessarily repetitious to include the name of the

area in the title. The title should be centered above the main body of the chart.

TYPES OF COLUMN CHARTS

For the sake of brevity as well as to avoid repetition, an effort has been made to limit the present discus-sion of column charts to basic design principles and specifications. Cursory reference is made to the simi-larities and differences between column and bar charts, mainly for the purpose of explaining the sig-nificance of certain structural components of the two types of charts.

The following paragraphs present a classification and description of eight different types of column charts, supplemented with graphic illustrations as de-picted in Figure 3-3.

Simple Column Chart.   In several of its basic fea-tures, the simple column chart has much in common with the simple bar chart. The base line of the column chart is drawn horizontally and under no circum-stances should it be omitted. The simple column chart is particularly valuable for showing time series.

Connected Column Chart.   This type of chart possesses characteristics of both the simple column chart and staircase surface chart. Although all the columns are distinct, there is no space between them. The connected-column chart may be particularly val-uable as a space-saving device.

Grouped Column Chart.   This chart is compara-ble to the grouped, multiple, or compound bar chart. Two, or occasionally three, columns representing dif-ferent series or different classes in the same series can be grouped together. In grouping the columns they may be joined together or separated by a narrow space.

Subdivided Column Chart.   The subdivided col-umn chart, like the subdivided, segmented, or compo-nent bar chart, is used to show a series of values with respect to their component parts. The subdivided col-umn chart is also similar to the subdivided surface chart. Ordinarily cross-hatching is used to differenti-ate the various subdivisions of the columns. The scale may be expressed in terms of absolute or relative values.

Net Deviation Column Chart.   The net deviation and the gross deviation column charts are similar to the bilateral bar charts. They emphasize positive and

---

[2] Department of the Army, *Standards of Statistical Presen-tation,* Department of the Army Pamphlet 325-10, 1966, p. 88.

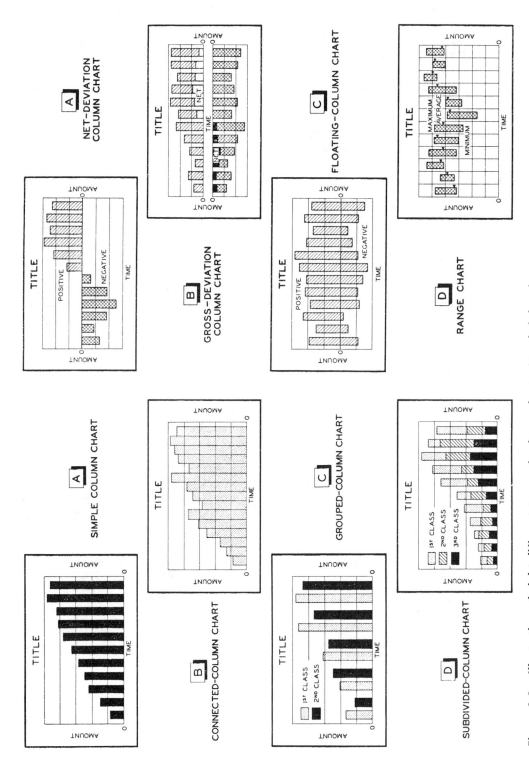

Figure 3-3. Illustration of eight different types of column charts. A verbal description of each type of column chart is included in the text. (From Calvin F. Schmid and Stanton E. Schmid, Handbook of Graphic Presentation, New York: John Wiley & Sons, 1979, pp. 80–81.)

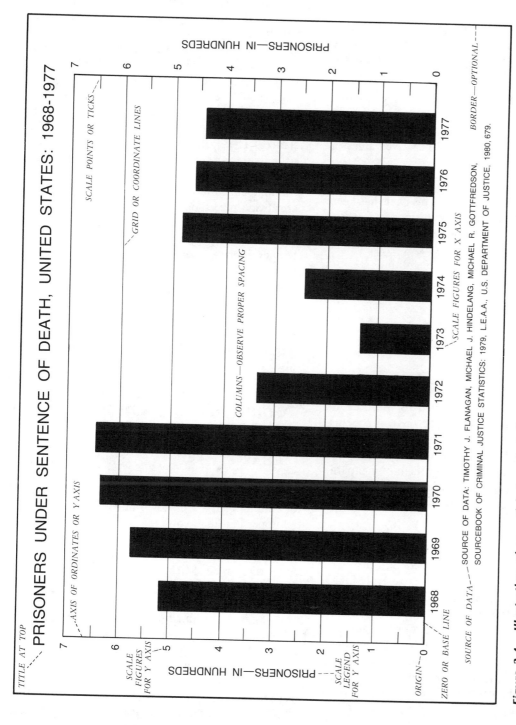

Figure 3-4. *Illustration of essential structural elements of a simple column chart. Design standards and specifications for various types of column charts are covered in the text.*

negative numbers, increases and decreases, and gains and losses. In the net deviation chart, the column extends either above or below the referent line, but not in both directions.

Gross Deviation Column Chart. The columns in this type of chart extend in both directions from the referent line. By means of cross-hatching, both gross and net changes can be readily portrayed.

Floating Column Chart. The floating column chart is a deviational or bilateral chart with 100 percent component columns. The deviations from the referent line represent positive and negative values or differential attributes.

Range Chart. The range chart shows maximal and minimal value in time series. This chart has been referred to as a "stock-price" chart, since it is extensively used in plotting highest and lowest daily, weekly, or monthly stock quotations. Average values also can be readily indicated on the columns.[3]

BASIC STRUCTURAL CHARACTERISTICS AND DESIGN SPECIFICATIONS FOR COLUMN CHARTS

Figure 3-4 represents a typical model of a simple column chart that has been designed to illustrate its characteristic structural elements or components. Since most of these elements are the same or similar to those included in the discussion of bar charts, it would be mere repetition to cover them in comparable detail. Accordingly, only brief explanatory and supplementary comments pertaining to the structure of column charts will be presented.

1. Columns. Much of what was said concerning bars is applicable to columns. As with bars in a bar chart, the columns should not be excessively long or short or disproportionately wide or narrow. The interspaces should be from about one-fourth to the full width of a column, depending on the size of the grid and the number of columns. Sometimes the columns may touch. When irregular time intervals occur, adjustments should be made in the interspacing of the columns. The ordering of the columns for a time series is chronological, begin-

ning with the earliest date on the left side of the grid.

2. Scale. The vertical scale of the typical column chart conforms to that of a rectilinear coordinate grid. The vertical scale represents size or amount and runs from bottom to top. There is always a zero base line that is made slightly heavier than the other horizontal scale lines. The vertical scale should never be broken. As indicated in the discussion of bar charts, it is permissible to break a single "freak" column under certain circumstances. The specifications and characteristics of scale figures and scale legends are similar to those required for the horizontal scale in a bar scale. In a column chart, the vertical scale figures and legend are always placed on the left side of the grid. However, if the chart is very wide, the scale figures and legend may be placed on the right side as well. For temporal series, each column or group of columns carries its own time label located directly underneath the zero base line.

The specifications and suggestions pertaining to shading and cross-hatching, stubs, and title that were included in the discussion of bar charts are also generally applicable to column charts.

ADDITIONAL SUGGESTIONS AND REQUIREMENTS FOR DESIGNING COLUMN AND BAR CHARTS

Certain significant design specifications and problems for column charts are shown in Figure 3-5. First, in the upper left-hand corner (1), there are illustrations of four standard grid forms. In the opinion of the author, form (b) is the most acceptable. Second, the upper right-hand sketches (2) indicate how the width of columns can influence the quality of a column chart. Third, the examples in group number (3) illustrate (a) an acceptable method of breaking a "freak" column, (b) spacing of time scale when intervals are unequal, and (c) an appropriate technique to indicate data are missing. Fourth, these two sketches (4) show how the overall proportions of a column chart can affect the visual impression a chart conveys as well as its effectiveness as a communication device. Fifth, the examples in group (5) further demonstrate how important it is to design the width of columns and the spacing of columns properly. Sixth, this illustration (6) shows that connected columns are preferable to a

[3] Calvin F. Schmid and Stanton E. Schmid, *Handbook of Graphic Presentation,* New York: John Wiley & Sons, 1979, pp. 61–62.

Figure 3-5. Standards and techniques for improving the design of column charts. (From American National Standards Committee Y15, Time-Series Charts, New York: The American Society of Mechanical Engineers, 1979, pp. 48–51.)

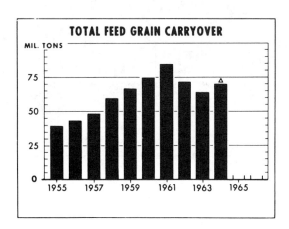

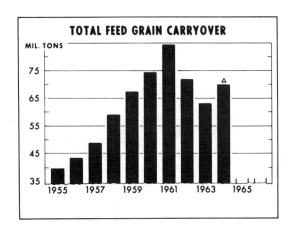

Figure 3-6. Column charts with and without zero referent line. The amputation of columns or bars by eliminating the zero referent line is unacceptable. Such a practice destroys the one-dimensional character of the column and bar chart. (From U.S. Department of Agriculture, Tips on Preparing Chart Roughs, Washington, D.C.: Office of Management and Services, 1973, p. 11.)

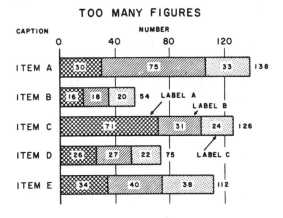

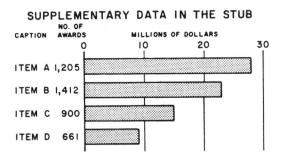

Figure 3-7. This chart shows the right and wrong way of adding supplementary or supporting data to a bar chart. As a strongly recommended practice, no statistical data should be superimposed in the body of the chart, particularly inside the bars. Statistical data may be placed to the left of the zero referent line as shown in the lower chart. [From National Institutes of Health, Division of Research Grants, Manual of Statistical Presentation, DRG Statistical Items, No. 10 (January 1970), p. 23.]

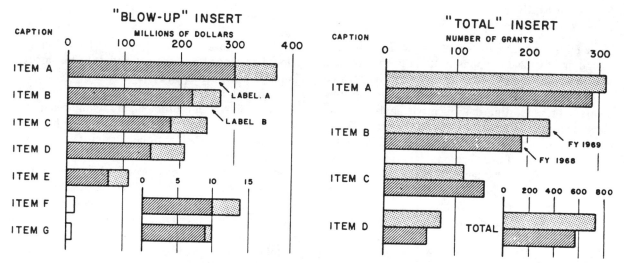

Figure 3-8. A simple bar chart with a "blow-up" insert and a grouped bar chart with a "total" insert. Inserts may be added to various graphic forms to round out, extend, or clarify certain information not shown by the primary chart. [From National Institutes of Health, Division of Research Grants, Manual of Statistical Presentation, DRG Statistical Items, No. 10 (January 1970), p. 17.]

series of relatively long, narrow separate columns. Seventh, the importance of selecting shading and cross-hatching patterns with care and discrimination is well illustrated by the examples in the two groups of columns (7) and (8) at the bottom of the page.

Figure 3-6 illustrates the serious consequences of the broken scale in column and bar charts. The upper chart has been designed correctly, and the lower chart, without a zero referent line, incorrectly. The

lower chart is untruthful and misleading. It will be seen that an amount representing 35 million tons out of a total of approximately 85 million tons has been eliminated from the vertical scale. As a result, the height of the column for 1955 is about one-third as large as it should be, and the basic comparability of the chart as a whole has been destroyed.

The component or subdivided bar chart in Figure 3-7 shows how cluttered and confusing a chart can be

Figure 3-9. Column chart with overlapping columns. Where an overlapping design is used for a column or bar chart it is particularly advantageous if the respective size of each set of columns or bars is consistently different, that is, if all, or at least most, of the companion columns or bars for each set are consistently larger or smaller. [From National Institutes of Health, Division of Research Grants, Manual of Statistical Presentation, DRG Statistical Items, No. 10 (January 1970), p. 23.]

INCORRECTLY OVERLAPPED SERIES

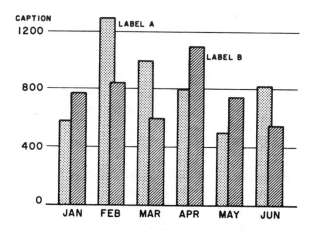

CORRECTLY OVERLAPPED SERIES

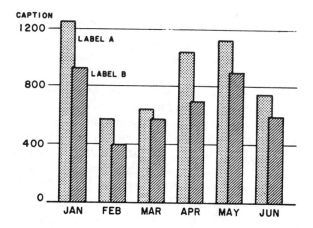

made by the superimposition of an excessive number of figures. It is readily apparent that the effectiveness of a chart as a communication device can be negated by this practice. If it is necessary to include supplementary or supporting data in a bar chart, it is acceptable to place them to the left of the zero referent line as shown in the simple bar chart in Figure 3-7.

For purposes of clarification, elaboration, or emphasis, it is often advantageous to add an insert to a larger chart. Figure 3-8 illustrates how this can be done on a bar chart. For example, in order to clarify and to supplement the composition of items F and G in the bar chart on the left side, an enlarged and detailed insert has been added. This is referred to as a "blow-up" insert. The grouped bar chart on the right side of Figure 3-8 has an insert showing the total comparative distribution for the two major categories. This type of insert is known as a "total" insert.

In a grouped column or bar chart, the columns or bars may be overlapped. This technique can save space as well as avoid very narrowly misproportioned columns and bars. However, in order to develop a successfully designed chart with overlapping columns or bars, it is important that one set of columns or bars be shorter in all, or virtually all, of the companion columns or bars in the other set. The chart on the left side of Figure 3-9 fails to meet this requirement; obviously it is not as satisfactory as the one on the right side.

Figure 3-10 was constructed to illustrate the inevitable consequences of failing to fulfill two indispensable requirements in the preparation of an acceptable statistical chart. The first requirement is to select the most appropriate type of chart as determined by the data, purpose at hand, and other essential considerations. The second requirement is to construct the chart in accordance with the highest standards and principles of chart design. The upper portion (A) of Figure 3-10 indicates failure in both respects. It reveals that a column chart is clearly not the most appropriate graphic form for portraying this particular series of data. Furthermore, as the chart has been laid out, it violates many principles and standards of good design. By contrast, the horizontal bar chart (B) with its simplicity and clarity is unquestionably superior both in terms of appropriateness and quality.

Some of the more obvious deficiencies and errors manifested by the chart in the upper part (A) are as follows: (1) It is overloaded with statistical data. (2) The fact that most of the figures are placed inside the bars (columns) adds to the depreciation of the chart. (3) There is no scale. (4) The labels for the various categories are misplaced. (5) The bars (columns) should be shaded or cross-hatched. (6) The bars (columns) are not arranged systematically.[4]

Figure 3-11 embodies helpful suggestions for resolving a few design problems in a simple bar chart when the stubs are long and the data slightly awkward for charting. The upper design (A) manifests two shortcomings: First, the stubs occupy too much space, and as a consequence the bars are crowded into a relatively small area. Second, besides causing an imbalance in the chart, the shortness of the bars makes them less discriminable. In order to resolve the problem of space and balance, two simple steps can be taken: First, instead of using one line for the stubs, the stubs can be placed on two lines. Second, since the longest bar extends only a short distance beyond the last scale line, the addition of another full-scale interval is superfluous and a waste of space. As a general rule, the addition of a full interval is not warranted unless the bar extends at least a half interval or more beyond the last scale interval. The logic and practicality of the two recommended adjustments are illustrated in the revised chart (B).

CRITIQUES OF SUBSTANDARD BAR AND COLUMN CHARTS: SOME LESSONS THAT CAN BE LEARNED

Two Simple Bar Charts

Both Figures 3-12 and 3-13 are examples of two simple bar charts that reflect several errors and deficiencies resulting from a failure to observe certain basic principles and standards of chart design. It is indeed puzzling to explain how it is possible to construct charts of this kind with so many shortcomings, especially since the requirements for a simple bar chart are so elementary and so easily understood.

With reference to Figure 3-12, there is no particular advantage or other pertinent reason for arranging the bars (columns) vertically on the page. It might be suggested that space was the determinant reason. However, the width and height of the chart measured from the base line are almost identical. The years

---

[4] Department of the Army, *Standards of Statistical Presentation,* Department of the Army Pamphlet 325-10, 1966, p. 88.

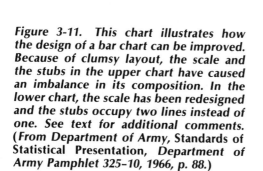

Figure 3-10. Illustration of proper and improper procedure and technique in selecting and constructing a statistical chart. The upper chart (A) indicates (1) incorrect selection of the most appropriate graphic form, and (2) violation of basic rules and standards of chart design. On the other hand, the lower chart (B) reflects acceptable procedure and technique. For further comments, see text. (From Department of Army, Standards of Statistical Presentation, Department of the Army Pamphlet 325-10, 1966 p. 65.)

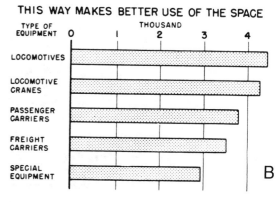

Figure 3-11. This chart illustrates how the design of a bar chart can be improved. Because of clumsy layout, the scale and the stubs in the upper chart have caused an imbalance in its composition. In the lower chart, the scale has been redesigned and the stubs occupy two lines instead of one. See text for additional comments. (From Department of Army, Standards of Statistical Presentation, Department of Army Pamphlet 325-10, 1966, p. 88.)

along the base line are superfluous and only serve to indicate the year for which the data are reported. These figures are mere clutter and could be readily replaced by a single explanatory note. As a horizontal bar chart, the stubs (names of countries) should be centered outside each bar. In comparison to the present arrangement, the horizontal lettering would be more readable. Moreover, it is poor practice to include virtually all of the lettering inside the bars.

With the bars set horizontally, the scale should be redesigned and scale lines added to make the chart more meaningful, more graphic, and easier to interpret.

Figure 3-13 is a simple bar chart of poor design and of questionable utility and significance as a medium of visual communication. The chart depicts the percentage distribution of opinions concerning performance ratings of police departments in terms of

Figure 3-12.   This chart exhibits several errors and deficiencies. The bars should be arranged horizontally rather than vertically; the years on the horizontal scale should be eliminated, and an explanatory note substituted; the location of the stubs is misplaced; and the scale should be redesigned and scale lines added (From National Commission on the Causes of Crime, Donald J. Múlvihill and Melvin M. Tumin (co-directors) and Lynn A. Curtis (assistant director), Crimes of Violence, *Vol. 11, Washington, D.C.: Government Printing Office, 1969, p. 129.)*

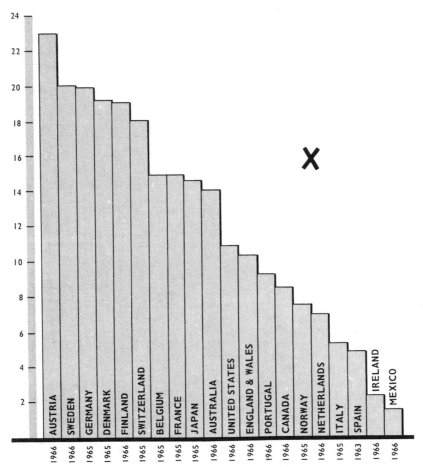

a Data are presented for last year available.
Source: United Nations, *Demographic Yearbook,* 1967, 19th edition.

Reported suicide rates for selected countries, 1965 or 1966. a *[rates per 100,000 population].*

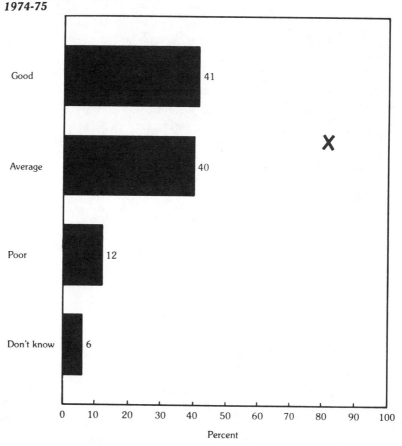

Residents of 26 central cities:
Ratings of police performance,
1974-75

Figure 3-13. A simple bar chart of inferior design. The
scale, bars, and overall composition of the chart are
strikingly incongruous. See text for specific comments.
(From Law Enforcement Assistance Administration,
United States Department of Justice, Myths and Reali-
ties About Crime, 1978, p. 9.)

Figure 3-14. A poorly planned and
poorly constructed column chart. The discrepancy between the vertical scale and
the plotting data is the most serious defect. (From Department of Justice, Immigration and Naturalization Service, 1975
Annual Report, Washington, D.C.: Government Printing Office, 1976, p. 23.)

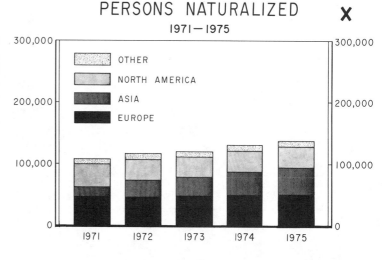

GROWTH OF THE MICROFILM COLLECTION
OF THE GENEALOGICAL SOCIETY OF UTAH

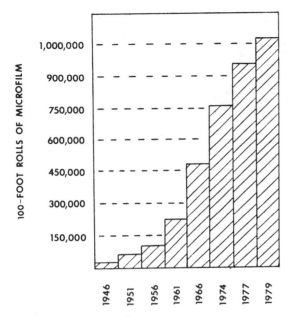

Source: The Genealogical Society of Utah
Dec. 31, 1979

Figure 3-15. In this chart, the spacing of the columns has not been adjusted to the unequal time intervals shown by the data. [From Office of Population Research, Princeton University and Population Association of America, Population Index, 46 *(1) (Spring 1980), front cover.]*

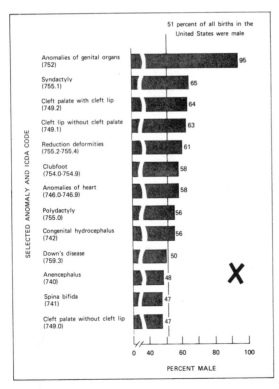

Percent male for births with selected anomalies, 1973-74 average.

Figure 3-16. The broken scale is contrary to the one-dimensional character of the bar and column chart. It is never an acceptable practice. Also, from a mechanical point of view, apparently little thought was given to the size of lettering in relation to the amount of reduction. [From Selma Taffel, Congenital Anomalies and Birth Injuries Among Live Births: United States, 1973-1974, *(DHEW Publication; No. (PHS) 79-1909, National Center for Health Statistics, Washington, D.C.: Government Printing Office, 1978, p. 7.]*

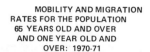

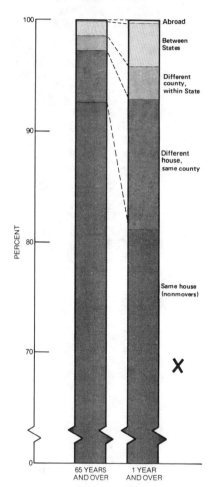

MOBILITY AND MIGRATION
RATES FOR THE POPULATION
65 YEARS OLD AND OVER
AND ONE YEAR OLD AND
OVER: 1970-71

Figure 3-17. This is another example of the broken scale as applied to a column chart. The confusing shading scheme on this chart further lowers its quality. [*From Jacob S. Siegel,* Current Population Reports, *Demographic Aspects of Aging and the Older Population in the United States,* Special Studies, Series P-23, No. 59, *United States Bureau of the Census, 1976, p. 20.*]

four categories—"good," "average," "poor," and "don't know." Assuming that it would be advantageous to chart a series of data of this kind, it is apparent that a much more effective and appealing design could have been devised. Just why an oversized, squarish grid with a scale extending to 100 percent was made the basic framework for a bar chart of this kind is difficult to understand. As a result, less than one half of the grid, including the scale, is utilized, and in order to fill in the space vertically, the bars and interspace were made disproportionately wide. If this chart were to be reconstructed, the following features should be incorporated. (1) A properly designed scale extending to 40 percent should be placed at the top of the chart. (2) The four bars should be less than one-half the width of the present bars and the interspaces about three-fifths the width of the new bars. (3) The

new bars should be about twice the length of the present ones.

A Column Chart with an Aberrant Vertical Scale

The immediate impression that one may derive from Figure 3-14 is the strange incompatibility between the grid and the five columns. The vertical scale extends to 300,000, but the maximal value of the data is less than 150,000. In other words, over half of the grid is superfluous, and the columns by contrast are disproportionately short. With the same sized grid and recalibrated scale intervals, a chart much better with respect to both detail and balance could be reconstructed. In addition, of course, improved shading,

State Government Intergovernmental Expenditure as a Percentage of Total State General Expenditures: 1973

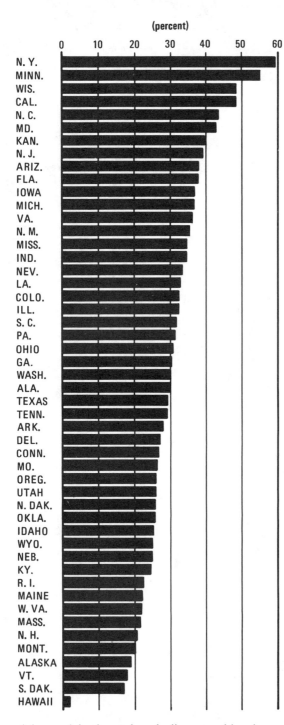

(percent)

Figure 3-18. This simple bar chart with 50 categories arranged in rank order fulfills the specifications of a well-designed chart. It is clear, neat, simple, reliable, and readily interpretable. (From U.S. Bureau of the Census, Chart Book of Governmental Data: Organization, Finance and Employment: 1973, Series GF 73, No. 7, Washington, D.C.: Government Printing Office, 1975, p. 10.)

ticks, and horizontal scale lines would enhance the quality of the chart.

A Column Chart with a Deficient Horizontal Scale

This chapter emphasizes that in portraying time series both the vertical and horizontal scales of column charts should be designed in accordance with accepted principles and practice. Figure 3-15 falls short of fulfilling this requirement, by disregarding the varying length of the time intervals in the plotting data. An examination of Figure 3-15 will reveal that the first five columns (1946–1966) are five years apart, while the sixth column (1974) follows eight years later. In the seventh column (1977), three years intervene and in the last column (1979) there is an interval

AVERAGE PAYMENT PER RECIPIENT FOR OLD AGE ASSISTANCE AND AID TO FAMILIES WITH DEPENDENT CHILDREN, INCLUDING VENDOR PAYMENTS FOR MEDICAL CARE, BY STATE, OCTOBER 1965.

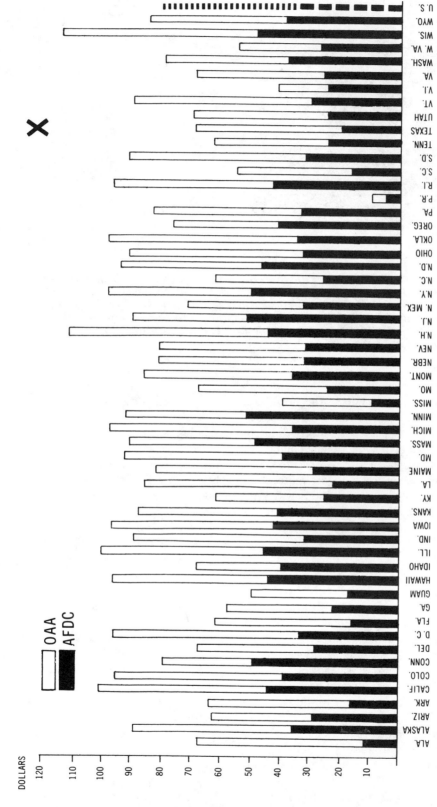

Figure 3-19.   This chart displays comparative data for 50 states and certain territories by means of a column (bar) chart. The columns (bars) are arranged alphabetically. Visually and statistically, the chart would be rated unacceptable. (From Department of Health, Education, and Welfare, Welfare Administration, Social Development, Washington, D.C.: Government Printing Office, 1966, p. 46.)

Prevalence of NFP (Including Rhythm) and Total Contraceptive Prevalence
Among Women of Reproductive Age in Union

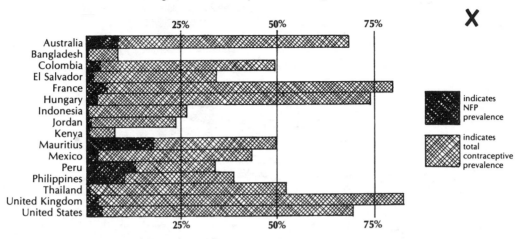

Source: Population Reference Bureau

Figure 3-20.   A poorly designed chart in which the bars are arranged alphabetically. In a chart of this kind, visual comparisons are extremely difficult to make. In addition to the deficiencies in basic design, the draftsmanship and other mechanical features of this chart are of mediocre quality. [From Population Crisis Committee, "Natural Family Planning: Periodic Abstinence as a Method of Fertility Control," Population (June 1981), 3.]

of two years. Some indication of the different time intervals should have been made in the horizontal scale. Of course, an alternative solution to a problem of this kind would be to select an entirely different graphic form to portray the data. Incidentally, an error in the vertical scale can also be observed. The first six intervals of the vertical scale are represented by units of 150,000, but the seventh interval is 100,-000—900,000 to 1,000,000—although the spacing is the same as for the first six intervals.

The Broken Scale and Its Consequences

Figure 3-16 represents a flagrant violation of the rationale of the one-dimensional chart by the continuous break in its scale. As a consequence, valid comparisons among the various categories are not possible. In light of what has already been said concerning this error, further comments would be superfluous.

With reference to this particular chart, it should be noted that the lettering is unusually small and difficult to read as a result of overreduction in the printing process.

Another example of the broken scale and its im-

plications are illustrated by the column chart in Figure 3-17. In addition to the broken scale, the chart is further depreciated by the inferior quality of the shading.

Serious as the broken scale is, there is a similar breach of principle that is just as serious, besides being more difficult to detect. This particular practice involves the amputation of columns without indicating a break in the scale. It is identical to the practice encountered so frequently in the design of rectilinear coordinate line charts, where the zero base line is nonexistent in an unbroken ordinal scale. Years ago (1923) Karl Karsten comments:

This practice of omitting the zero line is all too common, but it is not for that reason excusable. The amputated chart is a deceptive one.... A curve-chart without a zero line is in general no whit less of a printed lie, than a vertical bar chart (column chart) in which the lower parts of the bars themselves are cut away. The representation of comparative sizes has been distorted and the fluctuations (changes in values) exaggerated.[5]

---

[5] Karl G. Karsten, *Charts and Graphs,* New York: Prentice-Hall, 1923, pp. 155–156.

Bars and Columns Should Be Arranged Systematically: A Clarification

As indicated earlier, bars and columns should be arranged in some systematic order. For bar charts, the most common and most visually effective arrangement is according to size or value. Alphabetic and geographic arrangement may be classified as system-atic, but, from a visual standpoint, these are usually ineffective and incomprehensible. Figure 3-18 shows the percentage distributions of state government intergovernmental expenditures for the 50 states arranged in rank order. The scale, scale lines, bars, spacing, and other features of the chart facilitate comparisons among the states both from statistical and visual points of view. By contrast, the bars (columns) in Figure 3-19 are arranged alphabetically.

Figure 3-21.   This chart was prepared for the purpose of correcting the inadequacies and limitations, both in design and construction, of Figure 3-20. It will be noted that in order to make the chart more meaningful as well as to facilitate interpretation the basic design has been revised. Also, several mechanical features of the original chart have been changed.

PREVALENCE OF NFP (INCLUDING RHYTHM) AND TOTAL CONTRACEPTIVE PREVALENCE AMONG WOMEN OF REPRODUCTIVE AGE IN UNION

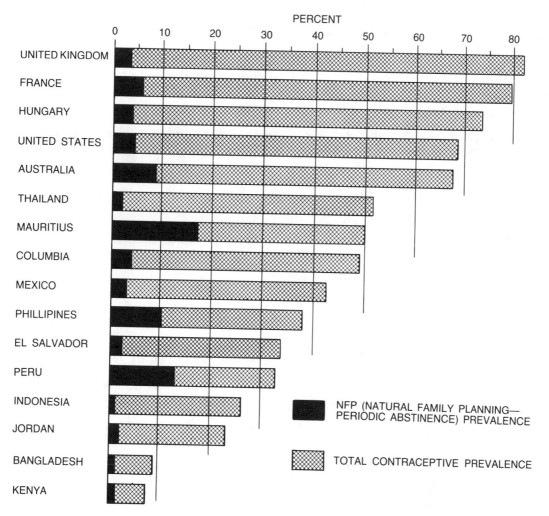

SOURCE: POPULATION REFERENCE BUREAU

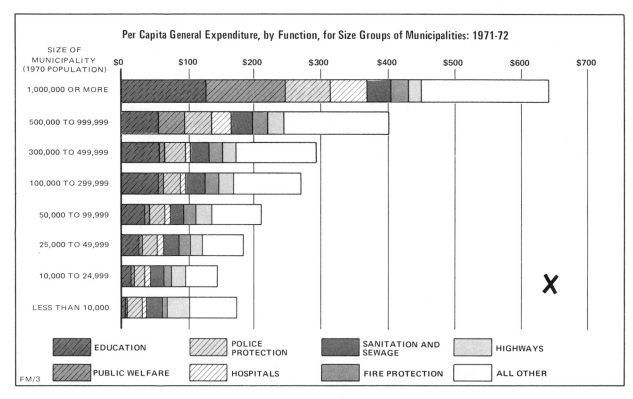

Per Capita General Expenditure, by Function, for Size Groups of Municipalities: 1971-72

SIZE OF MUNICIPALITY (1970 POPULATION)

EDUCATION		POLICE PROTECTION		SANITATION AND SEWAGE		HIGHWAYS	
PUBLIC WELFARE		HOSPITALS		FIRE PROTECTION		ALL OTHER	

FM/3

Figure 3-22. A component bar or column chart with eight divisions such as is found in this chart is extremely difficult, if not impossible to interpret. The inferior shading on the bars further depreciates the effectiveness of this chart. (From United States Bureau of Census, Topical Studies, Vol. 6, Graphic Summary of the 1972 Census of Governments, No. 5, Washington, D.C.: Government Printing Office, 1979, p. 79.)

Where We Ship Our Agricultural Exports

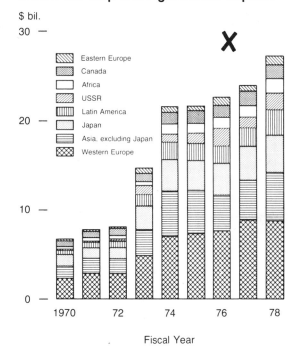

Figure 3-23. A column chart with eight components. This chart, like the preceding bar chart with eight components (Figure 3-22), cannot be interpreted with any degree of facility and clarity. Obviously this type of chart is inappropriate for the data that are being represented. (From United States Department of Agriculture, Handbook of Agricultural Charts, 1979, Agricultural Handbook No. 561, 1979, p. 79.)

The unit of comparison is dollars, representing average payments for old age assistance and aid to families with dependent children. As a medium of communication, both visually and statistically, it is confusing and ineffective. A properly arranged statistical table would be far more informative, reliable, and interpretable than a chart of this kind.

In a situation of this kind, one or the other of two choices should be made: First, if possible, redesign an effective chart. Second, abandon the preparation of a chart in favor of textual or tabular presentation.

As we have pointed out previously,

Before a decision is made to present data in graphic form there should be good and sufficient reasons for such a choice. The advantages should unmistakably outweigh any disadvantages in comparison to textual or tabular presentation. Charts

are not merely cosmetic appendages; they must serve a useful purpose based on careful insight and judgment.[6]

Again, Figure 3-20 shows how an alphabetic rather than a rank-order arrangement of bars can negate the logical basis and visual effectiveness of the direct simple comparison for which, ostensibly, a bar chart is designed. As Figure 3-20 now stands, it possesses more of the characteristics of a time-consuming puzzle than a clear, readily interpretable vehicle of visual communication. Using the same data, Figure 3-21 demonstrates most cogently the contrast in clarity and interpretability between alphabetically and rank-order arranged bar charts.

A Pitfall in Component Bar and Column Charts

A common shortcoming in component bar and column charts is the inclusion of too many divisions.

[6] Calvin F. Schmid and Stanton E. Schmid, *Handbook of Graphic Presentation,* New York: John Wiley & Sons, 1979, pp. 61–62.

Figure 3-24. *A column chart with two multiple-amount scales that has been constructed incorrectly. Note that one scale begins with zero and the other with 5000.* [*From University of Washington,* News Letter No. 25 (*October 15, 1938*) 1.]

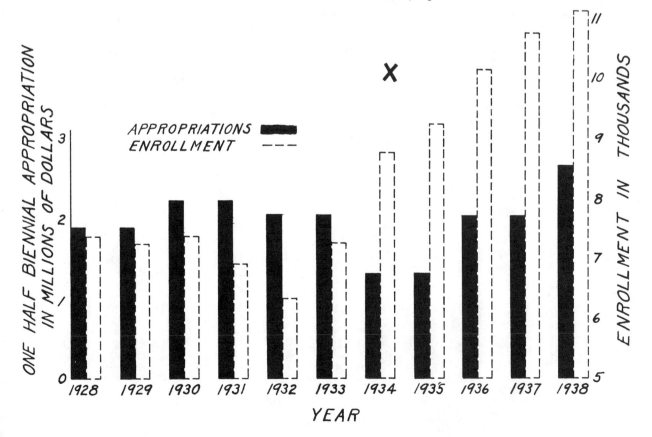

Such a practice may turn out to be futile and self-defeating since it is usually difficult, if not impossible, to make intelligent comparisons among the various components. Figure 3-22 illustrates a dilemma of this kind. There are eight different components in each bar representing per capita expenditures for education, public welfare, police protection, hospitals, sanitation and sewage, fire protection, highways, and so on. Very little reliable or meaningful information can be derived from this chart even after the most searching examination. The confusing and inadequate shading system contributes to the other deficiencies of the chart.

As a column chart, Figure 3-23 exhibits the same basic problem indicated in Figure 3-22. In each of the nine columns, there are eight different components showing the value and destination of United States agricultural exports. Although, in general, the chart shows good design features including cross-hatching, it is very difficult to interpret in a really meaningful sense. It seems clear that the component column chart is not the most appropriate graphic form to portray this series of data.

Column Charts with Multiple-Amount Scales

Column charts and other types of rectilinear coordinate time charts with two or more multiple-amount scales can be difficult to interpret, as well as being confusing. Because of these potential pitfalls, a few caveats are in order: (1) Charts of this kind should be constructed with meticulous care. (2) They should be used with discretion. (3) Multiple-amount scales normally should be limited to two scales. (4) The zero or

Figure 3-25. A reconstruction of Figure 3-24. Note that both scales begin with zero. As a result, the two sets of columns are comparable, and the discrepancies between them are less pronounced. Incidentally, the columns representing appropriations for 1928 and 1929 in Figure 3-24 have been plotted incorrectly. Also, besides exhibiting basic scale discrepancies, both the design and construction of Figure 3-24 are inferior.

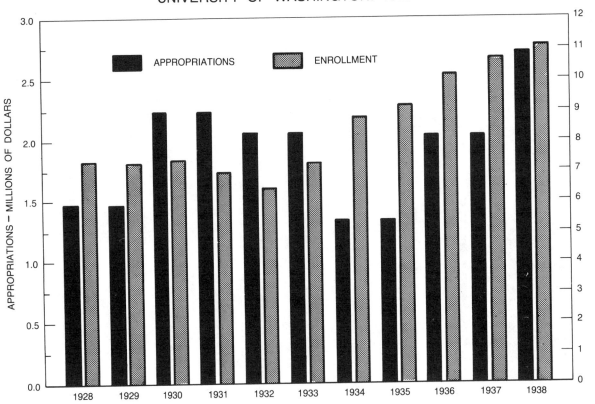

TRENDS IN APPROPRIATIONS AND ENROLLMENT
UNIVERSITY OF WASHINGTON: 1928-1938

MARKET HOGS AND PIG CROPS

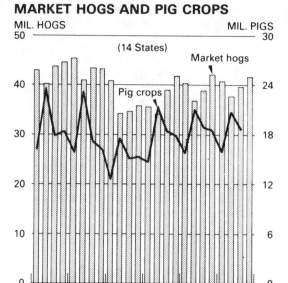

Pig crops — Dec.-Feb., Mar.-May, June-Aug., Sept.-Nov. Market hogs on farms —
Dec. 1 previous year,   March 1, June 1, Sept. 1. Dec. 1.

Figure 3-26. Another chart with two multiple-amount scales, the columns representing hog production and the curve, pig production. As indicated in the text, the question is raised concerning the need for a second scale. (From U.S. Department of Agriculture, 1978 Handbook of Agricultural Charts, Agriculture Handbook No. 551, Washington, D.C.: Government Printing Office, 1978, p. 109.)

COTTONSEED ACREAGE AND PRODUCTION

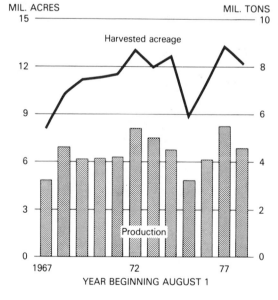

Figure 3-27. A combination of a column and a rectilinear coordinate line chart with two scales. One is expressed in millions of tons; the other, in millions of acres. Both scales begin with zero and are unbroken. See text for additional comments. (From United States Department of Agriculture, 1978 Handbook of Agricultural Charts, Agricultural Handbook No. 551, Washington, D.C.: Government Printing Office, 1978, p. 106.)

other base line should never be omitted. (5) The curves and/or columns should be gauged from a common base line, and the divisions of both scales should be spaced in the same manner.

Figure 3-24 is an example of a chart with two multiple-amount scales that has not been constructed correctly. The chart portrays trends in student enrollment and in legislative appropriations during an 11-year period, with particular emphasis on the message that the amount of money appropriated has been lagging behind the size of the student body. However, because of the erroneous design, the discrepancies among the two sets of columns are exaggerated. In this connection it will be noted that the scale for student enrollment begins with 500 and not zero. Figure 3-25 is a reconstruction of Figure 3-24, in which both variables begin with a common base line. The superiority of Figure 3-25 is clearly indicated. The columns still exhibit discrepancies, but they are based on reliable comparisons.

Figure 3-26 is primarily a column chart showing trends in hog and pig production. The columns represent the number of "market hogs" and the superimposed curve, the number of "pig crops." There are two scales, both expressed in millions: The one on the left indicates hogs, and the one on the right, pigs. Just why there are two scales on this chart when the primary scale on the left would suffice for plotting both series is difficult to explain. It would seem advantageous to use only one scale since it would simplify the structure of the chart as well as enhance its comparability and interpretability.

Unlike the preceding chart, the two scales in Figure 3-27 are expressed as two very different mensurational units, acres and tons. The data pertain to trends in cottonseed acreage and cottonseed production, which, as shown by the chart, reflect a noticeably high correlation. Although both series of data could have been plotted within the range of a single scale (left scale), there are sufficient reasons to use two scales. First, the basic units of the two series are entirely different. Second, if only one scale were used, the columns would be noticeably shorter and their respective values would not stand out as clearly. Third, the composition of the chart, especially the relation of the columns to the curve, would be less harmonious.

CHAPTER FOUR

MISCELLANEOUS GRAPHIC FORMS

Techniques and Standards

THE CHARTS SELECTED FOR INCLUSION IN this chapter possess special qualities and applications and thus are significantly different from those covered in other chapters of this book. Some consist of small families of a particular graphic form; others are independently distinctive or unique. In view of these facts, it would be neither logical nor appropriate to include them under any of the other major graphic types or special subjects discussed in other parts of this book. Accordingly, this chapter is devoted to design principles, issues, and problems as they relate to the following graphic forms: pie chart, simple frequency charts, cumulative frequency curves, Lorenz curve, charts portraying age-specific rates, scatter diagrams, and correlation matrices.

PIE CHART

Although the pros and cons concerning the utility and even the legitimacy of the pie chart have gone on for decades, the pie chart still occupies at least a tenuous place in the repertoires of most chart specialists.[1] Moreover, as a graphic form, the pie chart has always elicited much interest, besides enjoying wide popular and psychological appeal. A successful application of the pie chart as a vehicle of visual communication demands a clear understanding of what it can and cannot do. Certainly some of the criticism concerning the limitations and inadequacies of the pie chart is attributable to poor design. For example, Figure 4-1 is a pie chart with an extraordinarily large number of sectors, 33 in all. A pie chart with this many sectors is manifestly useless as a medium of visual communication. A simple, well-organized statistical tabulation would be much more satisfactory.

Figure 4-2, with 11 categories, is another example of a pie chart with too many sectors. This chart, with its well-chosen and well-executed hatching, lettering and overall composition, reflects superior draftsmanship, but the basic design is flawed by the excessive number of sectors. Is it possible to establish precisely the maximal number of sectors for a pie chart? In general, perhaps, a maximum of five or possibly six sectors would be acceptable. Five or less would be preferable.[2]

Figure 4-3 was designed to compare two series of data in which the respective size of each pie chart is of primary importance. However, examination of this illustration reveals that the circles are not comparable in size since they have been drawn incorrectly. It will be observed that during the fiscal year ending June 30, 1976, preference and nonpreference European immigrants admitted to the United States totaled 56,957 and Asiatic immigrants, 102,407. Since the ratio of European to Asiatic immigrants is 1.0 to 1.8, the designer of Figure 4-3 incorrectly laid out the diameters of the two circles on the basis of this ratio. By so doing, the area of the larger circle was made not 1.8 times as large, but 3.24 times as large. It is a simple mathematical principle that areas of circles vary as the square roots of their diameters. Accordingly, it is obvious that comparative pie charts or comparative circles in general based on direct ratios of diameters are erroneous and misleading. In such instances, the areas of the larger circles are exaggerated. Unfortunately, it may be difficult to detect distortions of this kind unless the data that were used in constructing the circles are clearly indicated. On the other hand, even where a series of pie charts is correctly constructed, interpretation of comparative size also may be found difficult. Not infrequently, in charts of this kind, there is a tendency to underestimate the com-

---

[1] The first of many psychological and psychophysical studies of various graphic forms conducted during a period of more than 50 years includes the pie chart. See Walter Crosby Eells, "The Relative Merits of Circles and Bars for Representing Component Parts," *Journal of the American Statistical Association,* **21** (1926), 119–132. For a more detailed review of similar studies, see Chapter 1.

[2] In laying out percentages in a pie chart, a percentage protractor can save time and effort. The percentage protractor was first suggested by Frederick E. Croxton, "A Percentage Protractor," *Journal of the American Statistical Association,* **17** (1922), 108–109.

Subject Content of the Harvard University Library

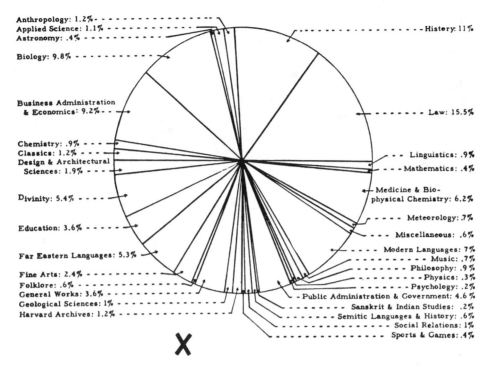

Anthropology: 1.2% - - - - - - - - - - - - - - - - -
Applied Science: 1.1% - - - - - - - - - - - - - - -
Astronomy: .4% - - - - - - - - - - - - - - - - - -
Biology: 9.8% - - - - - - - - - - - - -
Business Administration & Economics: 9.2% - - -
Chemistry: .9% - - - -
Classics: 1.2% - - - -
Design & Architectural Sciences: 1.9% - - -
Divinity: 5.4% - - - -
Education: 3.6% - - - - -
Far Eastern Languages: 5.3% -
Fine Arts: 2.4% - - - - - - - - -
Folklore: .6% - - - - - - - - - -
General Works: 3.6% - - - - - - - - - -
Geological Sciences: 1% - - - - - - - - - - - -
Harvard Archives: 1.2% - - - - - - - - - - - - -

History: 11%
Law: 15.5%
Linguistics: .9%
Mathematics: .4%
Medicine & Bio-physical Chemistry: 6.2%
Meteorology: .7%
Miscellaneous: .6%
Modern Languages: 7%
Music: .7%
Philosophy: .9%
Physics: .3%
Psychology: .2%
Public Administration & Government: 4.6%
Sanskrit & Indian Studies: .2%
Semitic Languages & History: .6%
Social Relations: 1%
Sports & Games: .4%

Figure 4-1. *A pie chart with 33 sectors. The clutter and confusion of this chart leaves much to be desired as a medium of visual communication. (From Keyes D. Metcalf,* Report on the Harvard University Library, *Cambridge, Mass.: Harvard University Library, 1955, p. 73.)*

parative size of larger circles. Rather than include several pie charts of varying size in a single illustration, it probably would be more expedient and effective to use a bar or column chart.[3]

It is common knowledge among graphic specialists that, in a practical sense, there is no single criterion for estimating values in a pie chart. Some users base estimates on the arcs of sectors, some on the areas of sectors, some on the central angles of sectors, and a very small number, on chords.[4] The variable and apparently uncertain criteria used in estimating values in pie charts contrast markedly with the more specific and accurate one-dimensional criterion of distance that characterizes bar and column charts.

Is it possible to eliminate the major contradictions and other deficiencies of the pie chart and still retain its basic characteristics? Olayinka Y. Balogun has sought ways to improve the pie chart by substituting the "decagraph,"

that epitomizes the good qualities of the pie graph and the other graphs.... The decagraph is an equal-sized decagon, with all vertices touching on a common circle, used as a graph to show component parts, which are usually expressed as percentages of the whole. The decagraph is basically similar to the pie graph. The only difference is the addition of chords and the removal of arcs.[5]

---

[3] For a more detailed discussion of two- and three-dimensional symbols, particularly as they pertain to statistical maps, see Chapter 7.

[4] In the 1926 study cited previously, Walter Crosby Eells found that in estimating the size of sectors, 51 percent of the subjects tested used arcs of sectors; 25 percent, areas; 23 percent, central angles; and 1 percent, chords.

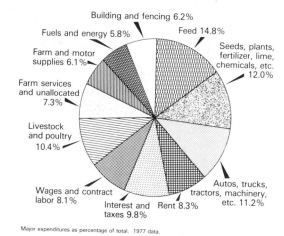

FARM PRODUCTION EXPENDITURES

Building and fencing 6.2%
Fuels and energy 5.8%
Farm and motor supplies 6.1%
Farm services and unallocated 7.3%
Livestock and poultry 10.4%
Wages and contract labor 8.1%
Interest and taxes 9.8%
Rent 8.3%
Autos, trucks, tractors, machinery, etc. 11.2%
Seeds, plants, fertilizer, lime, chemicals, etc. 12.0%
Feed 14.8%

Major expenditures as percentage of total. 1977 data.

Figure 4-2. Another example of a pie chart with two many sectors. (From U.S. Department of Agriculture, 1978 Handbook of Agricultural Charts, Agriculture Handbook No. 551, Washington, D.C.: Government Printing Office, 1978, p. 11.)

Figure 4-3. One of the basic objectives of this chart is to compare the number of European immigrants (56,957) with that of Asiatic immigrants (102,407). However, the areas of the two circles are not comparable, since they were laid out incorrectly. The ratio of 56,957 to 102,407 is 1:1.8, but the area of the larger circle is about 3.24 times as large as the smaller one. In constructing the circles, the diameters rather than the square roots of the diameters were taken as the basis of comparison. (From Department of Justice, Immigration and Naturalization Service, 1976 Annual Report, Washington, D.C.: Government Printing Office, 1978, p. 7.)

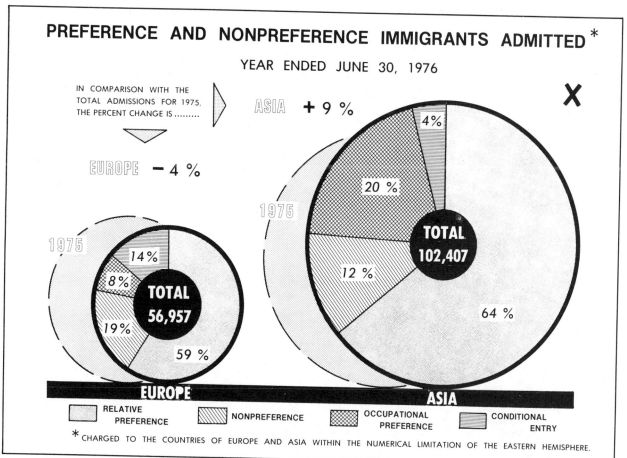

PREFERENCE AND NONPREFERENCE IMMIGRANTS ADMITTED *

YEAR ENDED JUNE 30, 1976

IN COMPARISON WITH THE TOTAL ADMISSIONS FOR 1975, THE PERCENT CHANGE IS ....

ASIA + 9 %

EUROPE − 4 %

1975

1975

TOTAL 56,957

14%
8%
19%
59 %

TOTAL 102,407

4%
20 %
12 %
64 %

EUROPE

ASIA

RELATIVE PREFERENCE

NONPREFERENCE

OCCUPATIONAL PREFERENCE

CONDITIONAL ENTRY

* CHARGED TO THE COUNTRIES OF EUROPE AND ASIA WITHIN THE NUMERICAL LIMITATION OF THE EASTERN HEMISPHERE.

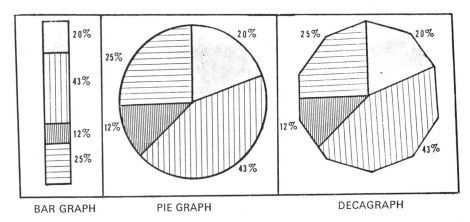

BAR GRAPH   PIE GRAPH   DECAGRAPH

Figure 4-4.   A comparison of the "decagraph" with the pie chart and the 100-percent column chart. Will the "decagraph" eventually supplant the pie chart? [From Olayinka Y. Balogun, "The Decagraph: A Substitute for the Pie Graph?" The Cartographic Journal, 15 (1978), 78–85.]

(See Figure 4-4.) In order to evaluate the comparative effectiveness of the pie chart and the decagraph on rapidity of estimation and accuracy of judgment, Balogun conducted detailed tests. In light of these tests, the decagraph was clearly superior to the pie chart. Specifically, these tests showed that the side of the decagon, the only new element added to the pie chart, was of crucial significance as the preferred basis for estimating values. According to Balogun, this finding "is indisputable evidence that the map users wanted another means of estimating apart from the arc, the angle, and the area." In addition, the decagraph elicited strong psychological appeal: 88.9 percent preferred the decagraph, 6.3 percent preferred the pie chart, and 4.8 percent were indifferent.

SIMPLE FREQUENCY CHARTS

Constructing a Frequency Distribution

One of the first steps in constructing a frequency distribution is to determine the most appropriate size and number of class intervals. A frequency distribution classifies, condenses and simplifies data for the purpose of analysis, interpretation, and presentation. If the size of class intervals is too small, there may be too many intervals, some of which will have relatively few cases or none at all. On the other hand, if the size

of the intervals is relatively large, there may be too few intervals and the essential characteristics of the data will be concealed. If the class intervals are not well chosen, the advantages of convenient summarization will be lost, and analysis and presentation may be more difficult. In actual practice, it is not possible to formulate precise rules concerning the most appropriate size or the optimal number of class intervals.

In order to illustrate how the size of class intervals actually influences frequency distributions, the reader is referred to Figure 4-5. The data shown on the chart represent the age distribution of 643 females who attempted suicide in the city of Seattle during the five-year period, 1948–1952. For example, in Panel A intervals of 1 year are shown; at the other extreme, in Panel F, the size of the intervals is 20 years. Incidentally, there are over 70 intervals in the distribution in Panel A as compared to 4 intervals in the distribution in Panel F. Among the six different sizes of class intervals, which one reveals the most characteristic and significant features of the distribution and summarizes the date most effectively for analysis and presentation? The choice would seem to be between the distribution in either Panel C or D. Since five years is larger, more manipulatable and more commonly used as a class interval, the distribution in Panel D would be the most satisfactory choice.

Three Simple Frequency Charts

Common techniques for the graphic portrayal of frequency distributions include (1) the histogram, (2) the polygon, and (3) the smoothed frequency curve.

---

[5] Olayinka Y. Balogun, "The Decagraph: A Substitute for the Pie Graph?" *The Cartographic Journal,* **15** (1978), 78–85.

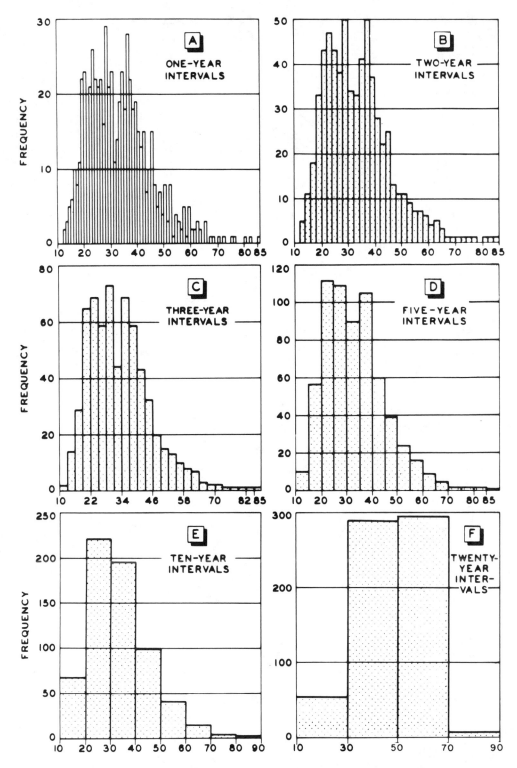

Figure 4-5. This chart illustrates how very important the size of class intervals is in designing a frequency distribution. (From Sanford M. Dornbusch and Calvin F. Schmid, A Primer of Social Statistics, New York: McGraw-Hill, 1955, p. 17.)

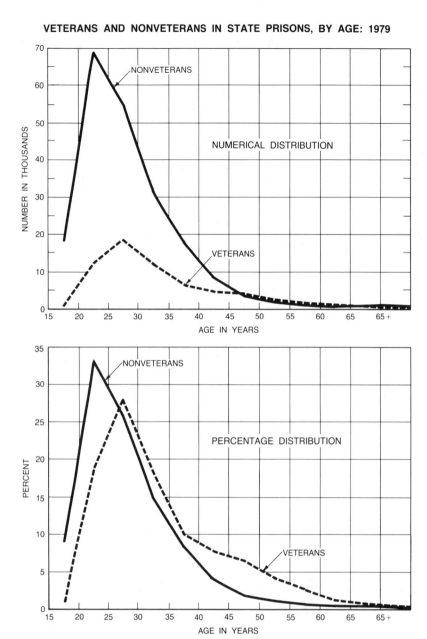

VETERANS AND NONVETERANS IN STATE PRISONS, BY AGE: 1979

Figure 4-6.   *Examples of frequency polygons. One set of polygons is plotted in absolute numbers; the other set, in percentages. Note the location of the plotting points. (Chart based on data furnished by Carol B. Kalish and Margaret Heisler of the U.S. Bureau of Justice Statistics.)*

These charts are frequently referred to as simple frequency graphs and are constructed on a rectilinear coordinate grid with class intervals represented on the abscissal axis and frequencies on the ordinal axis. They are visual models that enable one to comprehend essential features of frequency distributions; also they are invaluable tools for facilitating analyses and deriving solutions to statistical problems.

The typical histogram is constructed by erecting vertical lines at the limits of the class intervals as the basic step in forming a series of contiguous rectangles or columns. The six charts in Figure 4-5 are examples

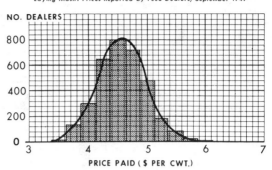

FREQUENCY DISTRIBUTION

Laying Mash: Prices Reported by Feed Dealers, September 1949

Figure 4-7. The basic graphic form of this simple frequency distribution is a histogram on which a smoothed curve has been superimposed. The smoothed curve closely typifies what is generally referred to as a "normal distribution." The histogram is based on sample data. (From Frederick V. Waugh, Graphic Analysis in Agricultural Economics, Washington, D.C.: U.S. Department of Agriculture, Agricultural Marketing Service, 1957, p. 3.)

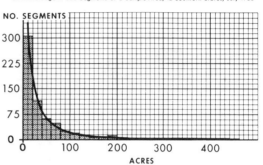

FREQUENCY DISTRIBUTION

Corn: Acreage in 623 Segments of a Sample Area, 12 Southern States, July 1956

Figure 4-8. Another illustration of a smoothed frequency distribution. In comparison with Figure 4-7, this distribution is characterized by pronounced skewness. This chart also is based on sample data. (From Frederick V. Waugh, Graphic Analysis in Agricultural Economics, Washington, D.C.: U.S. Department of Agriculture, Agricultural Marketing Service, 1957, p. 5.)

of histograms. In each of the three simple frequency charts—histogram, polygon, and smoothed frequency curve—there is a zero point on the ordinal scale, but not necessarily, or even usually, on the abscissal scale. In a histogram, the areas above the various intervals are proportional to the respective frequencies in the intervals.

In constructing a frequency polygon, the horizontal and vertical scales are the same as in a histogram. The appropriate frequency of each class is located at the midpoint of the interval, and the plotting points are connected by straight lines, thus forming a polygon. Sometimes polygons are constructed in which abscissal lines represent specific values; in such cases, the plotting of frequencies is located on the designated lines. Generally, in a frequency polygon it is assumed that the number of cases in each interval is concentrated at the midpoint of the interval, whereas in a histogram it is assumed that there is a uniform distribution of cases within each interval. If the same series of data is plotted as both a histogram and a polygon, certain significant similarities and differences in the two charts will occur. The total area under each chart will be equal. Both the histogram

and the polygon represent frequency, but unlike the columns in a histogram, the areas for specific intervals in a polygon are not proportional to the frequencies. For this reason, histograms are considered to be a more satisfactory graphic representation of frequency distributions than polygons. The frequency polygon is especially useful in portraying two or more series of data on the same chart, since polygons are more clearly distinguishable than histograms, which are superimposed on one another (Figure 4-6). As will be observed, frequencies may be expressed in absolute numbers or in percentages.

The smoothing of a frequency distribution can turn out to be a precarious, uncertain, and difficult undertaking. It should be done very selectively and with extreme care. In smoothing a histogram or polygon, the primary objective is to iron out or eliminate accidental irregularities resulting from what is presumed to be due to sampling errors. A smoothed curve is designed to depict an idealized configuration of a distribution of the population or universe from which the sample is derived. Although the original chart is constructed from observations based on a sample, the smoothed curve represents certain esti-

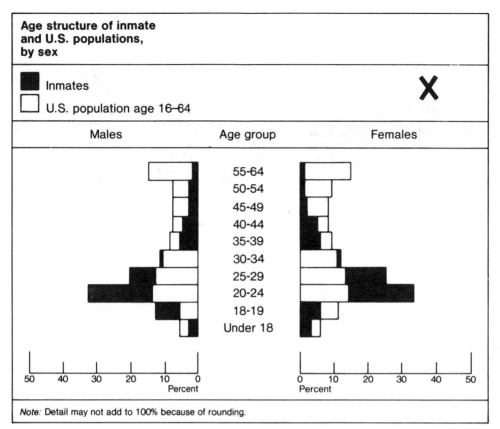

Figure 4-9. *This chart consists of four histograms. It was designed to compare the age and sex structure of local jail inmates throughout the country with that of the general population of the United States. However, no adjustments were made in this chart for unequal class intervals. As a result, the chart is incorrect and misleading. (From U.S. Department of Justice, Bureau of Justice Statistics,* Profile of Jail Inmates, *Washington, D.C.: Government Printing Office, 1980, p. 3.)*

mates and inferences derived largely from these data. The results of the smoothing process can turn out to be very uncertain since it may not be possible to determine whether particular humps, depressions, or other irregularities in a curve are the results of sampling errors or are indicative of some meaningful trend. If a frequency distribution is smoothed, it is suggested that the observed histogram or at least the observed plotting points of a polygon be included with the curve.

Figures 4-7 and 4-8 are examples of two histograms with smoothed frequency curves superimposed. These distributions clearly manifest two distinctly different configurations. Figure 4-7 approximates what is customarily referred to as a "normal distribution curve." On the other hand, Figure 4-8 is extremely skewed.

The columns of the histogram in Figure 4-7 indicate the number of dealers reporting prices paid by farmers for laying mash, and the smoothed curve an estimate or judgment concerning the general nature of the observed distribution. In smoothing the curve, the author states that

I have drawn a smooth curve representing a judgment as to the general nature of the observed distribution. Note that I have not bent the curve to make it go through the midpoint of each bar. Rather, I have drawn it smooth, to show the general nature of the distribution.[6]

[6] Frederick V. Waugh, *Graphic Analysis in Agricultural Economics,* Washington, D.C.: U.S. Department of Agriculture, Agricultural Marketing Service, 1957, p. 2.

It must not be assumed, of course, that curve fitting is always an intuitive arbitrary, trial-and-error procedure. Various mathematical techniques, some relatively simple, others complex, are also utilized.

Figure 4-8 summarizes the results of a survey of corn acreage made in 12 southern states based on a sample of 623 segments. The height of the columns of the histogram indicates the number of segments reporting corn acreage for the various class intervals. The smoothed curve is a generalized estimate of the distribution of corn acreage in the southern states. The fluctuations reflected in the chart may be the result of sampling errors.[7]

Designing Histograms with Unequal Class Intervals

A common error in constructing histograms with unequal class intervals is to make the heights of the respective rectangles commensurate with the frequencies of the class intervals without taking into consideration the width of the class intervals. For example, see Figure 4-9. It will be observed that 4-9 is based on four frequency distributions. Furthermore, we hasten to point out that although the positions of the abscissal and ordinal axes are reversed as in a conventional age and sex pyramid, Figure 4-9 is not an age and sex pyramid. The four histograms are based on four sex and age distributions that are en-

[7] Most graphic specialists would concur that the quality of Figures 4-7 and 4-8 and Figures 4-14 and 4-15 has been depreciated by the superfluity of grid lines. The unnecessarily large number of grid lines is attributable to the utilization of preprinted cross-ruled paper in which photosensitive ink was used. Unfortunately the grid lines as well as the frequency curves were picked up by the camera in the reproduction process. This could have been avoided if light blue ink had been used for the cross-rulings.

Figure 4-10. The results of adjusting the four histograms in the preceding chart for class intervals of unequal size are presented in this chart. In order to facilitate comparison between Figures 4-9 and 4-10, the same basic type of chart was retained. The differences between the two charts are especially noticeable in the case of the two 2-year age intervals and the one 10-year interval.

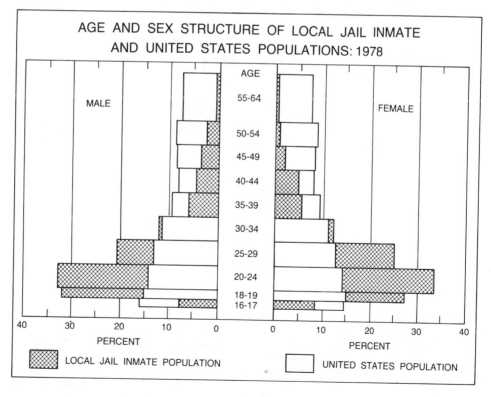

tirely separate statistically. There are two histograms representing the population of the United States aged 16–64, and two histograms representing inmates of local jails located throughout the country. It will be observed that the age distributions for the histograms consists of seven 5-year intervals, two 2-year intervals, and one 10-year interval.[8]

In Figure 4-9, all the intervals are equal in width, but in a histogram the relative frequency and the area of each rectangle are proportional. That is, since area equals width times height, the height of each rectangle as well as its area is proportional to the frequency in the class interval it represents. If the class intervals are all of equal width, the width can be considered to be unity. In such instances, the height of the rectangle is simply equal to the frequency in the class interval. However, if the class intervals are of unequal width, the same basic principle applies, and the proper height is found by dividing the frequency by the width of each class interval. Any convenient interval may be chosen as unity, and the proper divisors can be computed from this interval. Using these divisors, the height for each rectangle of the histogram can be computed readily for any combination of class width. In redesigning Figure 4-9, an interval of five was used as unity. The results of these adjustments are portrayed in Figures 4-10 and 4-11.

A comparison of Figure 4-9 with Figure 4-10 clearly reveals the errors and distortions that occur when histograms with unequal class intervals are improperly designed. It is obvious that two-year or ten-year class intervals should not be treated the same as the dominant five-year class intervals. Figure 4-9 violates both graphic and statistical principles.

It should be pointed out that in order not to make comparisons between Figures 4-9 and 4-10 unnecessarily complicated, no change was made in the basic design of the corrected chart. However, an alternative

[8] Figure 4-9 from U.S. Department of Justice, Bureau of Justice Statistics, *Profile of Jail Inmates,* Washington, D.C.: Government Printing Office, 1980, p. 3. Statistical transcripts furnished by Carol B. Kalish of the Bureau of Justice Statistics and John F. Wallerstedt of the Bureau of the Census. In the original tabulations the "under 18" interval includes ages 16–17.

Figure 4-11.   This chart is presented as a simpler and perhaps a more readily interpretable design than the design used in Figures 4-9 and 4-10.

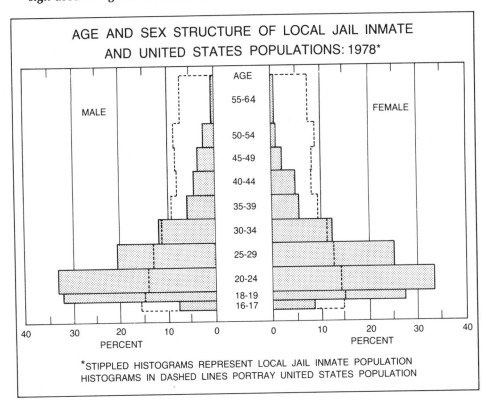

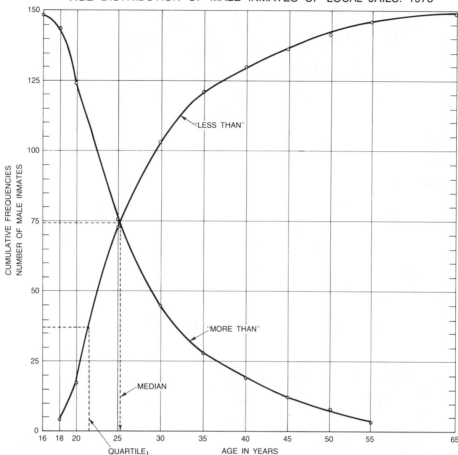

CUMULATIVE FREQUENCY CURVES OR OGIVES
AGE DISTRIBUTION OF MALE INMATES OF LOCAL JAILS: 1978

Figure 4-12. The essential properties of both "less than" and "more than" ogives are illustrated by this chart. Note especially the respective locations of plotting points, the cumulative frequencies on the vertical axis, and the interpolation lines and values for the median and first quartile. Chart based on data from special transcripts furnished by Carol B. Kalish of the Bureau of Justice Statistics and John F. Wallerstedt of the Bureau of the Census.

design, one that I consider simpler and more readable, is presented in Figure 4-11.

A Broken Vertical Scale in a Histogram is Never Justified

Another design practice that does violence to both graphic and statistical principles on which histograms are based is the broken scale. The ordinal scale that gauges frequencies should never be broken. As indicated previously, when the width of the rectangles in a histogram is laid out correctly, the various heights of the rectangles are directly proportional to the quantities charted. If the zero line is omitted or if the scale is broken, the chart ceases to be an acceptable histogram. The vertical dimension of the rectangles reflects a wrong visual impression, and the overall configuration of the distribution is seriously distorted.

CUMULATIVE FREQUENCY CURVES OR OGIVES

As indicated previously, in a simple frequency chart the number of cases or frequency for each class inter-

GRAPHIC REPRESENTATION OF CLASSROOM USE

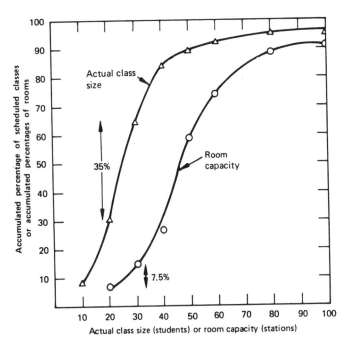

Figure 4-13. *Good examples of two "less than" ogives. They show the relationship between actual size and room capacity as part of a space utilization survey. See text for further comments. (From D. Kent Halstead,* Statewide Planning in Higher Educa-*tion,* Office of Education, U.S. Department of Health, Education and Welfare, Wash-*ington, D.C.: Government Printing Office, 1974, p. 453.)*

val is indicated separately. In a cumulative-frequency distribution the frequencies of successive class intervals are accumulated, beginning at either end of the distribution, and carried through the entire range of frequencies. If the accumulation is from the least to the greatest, it is referred to as a "less than" type of distribution; if from greatest to least, a "more than" type of distribution.

A useful type of chart including either or both a "less than" or "more than" curve can be constructed from a cumulative frequency distribution. Cumulative frequency curves of this kind are also called summation curves or ogives. Figure 4-12 is an illustration of both "less than" and "more than" ogives based on the age distribution of 148,003 male inmates of local jails. The data represent one of the series of data portrayed in Figures 4-9, 4-10, and 4-11. The vertical axis may represent either numerical cumulative frequencies as in Figure 4-12 or percentages, ranging from zero to 100 percent. Ogives can be used to determine as well as portray the number of proportion of cases above or below a given value. They also can be used for interpolating graphically the me-

dian, quartiles, deciles, or other measures of this kind.[9]

The most common design error in constructing either a "less than" or "more than" ogive is the selection of incorrect plotting points. In the "less than" curve, the correct plotting points are located at the upper boundary of each class interval; the "more than" curve, the plotting points are placed at the lower boundary of each class interval (see Figure 4-12).

Another useful application of the ogive which involves an interesting design problem is exemplified in Figure 4-13. The accumulated frequencies are expressed as percentages. One ogive shows scheduled classes; the other, room capacity. The problem relates to space utilization and optimal class scheduling. For example, if all stations in all rooms were occupied during scheduled classes, the two curves would be identical. To reduce the gap between the two curves

[9] Calvin F. Schmid and Stanton E. Schmid, *Handbook of Graphic Presentation,* New York: John Wiley & Sons, 1979, pp. 134–137.

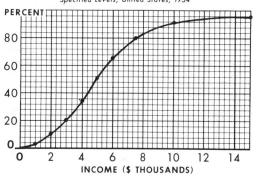

CUMULATIVE FREQUENCY
Percentage of Families With Incomes Below
Specified Levels, United States, 1954

Figure 4-14. *A common use of the ogive or summation curve. The chart portrays the distribution of income among American families in 1954. Note the unequal class intervals and the smoothness and symmetry of the curve. (From Frederick V. Waugh,* Graphic Analysis in Agricultural Economics, *Washington, D.C.: U.S. Department of Agriculture, Agricultural Marketing Service, 1957, p. 7.)*

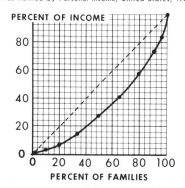

LORENZ CURVE
Families Ranked by Personal Income, United States, 1954

Figure 4-15. *The Lorenz curve is always plotted in a square field with equal percentage units from 0 to 100 on each axis. It is obvious that if there were a consistently equal distribution of income for all of the families, the curve would be in the form of a diagonal straight line. However, this is not the case. For example, the "poorest" 50.6 percent of the families received 27.2 percent of the income. (From Frederick V. Waugh,* Graphic Analysis in Agricultural Economics, *Washington, D.C.: U.S. Department of Agriculture, Agricultural Marketing Service, 1957, p. 9.)*

Table 4-1

Percentage of Families with Personal Income Below Specified Amounts and Percentage of Total Personal Income Obtained by These Families, United States: 1954

Income Less Than (dollars)	Percentage of	
	Families	Income
1,000	2.7	0.2
2,000	9.9	2.1
3,000	20.0	6.4
4,000	34.2	14.8
5,000	50.6	27.2
6,000	65.3	40.7
7,500	79.9	57.0
10,000	91.4	73.3
15,000	96.5	83.7
∞	100.0	100.0

Source: Frederick V. Waugh, *Graphic Analysis in Agricultural Economics*, Washington, D.C.: U.S. Department of Agriculture, Agricultural Marketing Service, 1957, p. 9.

and improve utilization, it is necessary to identify at what point the greatest imbalances between class size and room capacity occur. Graphically, they occur where the slopes of the two curves are least similar. For example, 35 percent of the scheduled classes have an enrollment of 20 to 30 students, yet only 7.5 percent of the rooms contain this number of student stations. There are also too few rooms that can accommodate 30 to 40 students, and there is an excess of large rooms. Fifteen percent of the rooms seat from 50 to 60 students, while only 3 percent of the classes include this many students.[10]

One of the most common applications of cumulative frequency curves is the portrayal of the distribution of wages and income. For example, a "less than" ogive can show the number or percentage of individ-

[10] D. Kent Halstead, *Statewide Planning in Higher Education,* Office of Education, U.S. Department of Health, Education, and Welfare, Washington, D.C.: Government Printing Office, 1974, pp. 452–453. Chart adapted from State of Illinois Board of Higher Education, *State-wide Space Survey,* 1968, p. 42.

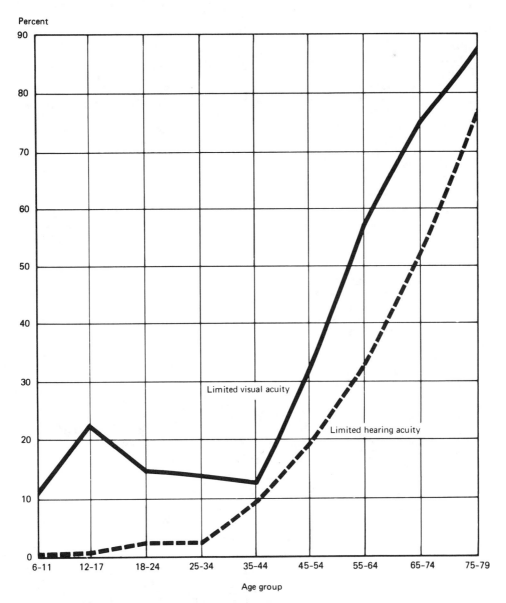

Figure 4-16. This chart was designed to portray the proportion of the population manifesting limited hearing and visual acuity according to nine age categories. The chart indicates that all nine age categories are of equal size, which actually is not the case. See Figure 4-17 for a corrected version of this chart. (From U.S. Department of Commerce, Office of Federal Statistical Policy and Standards, Social Indicators, 1976, Washington, D.C.: Government Printing Office, 1977, p. 176.)

PREVALENCE OF LIMITED HEARING AND VISUAL ACUITY, BY AGE: 1960-70

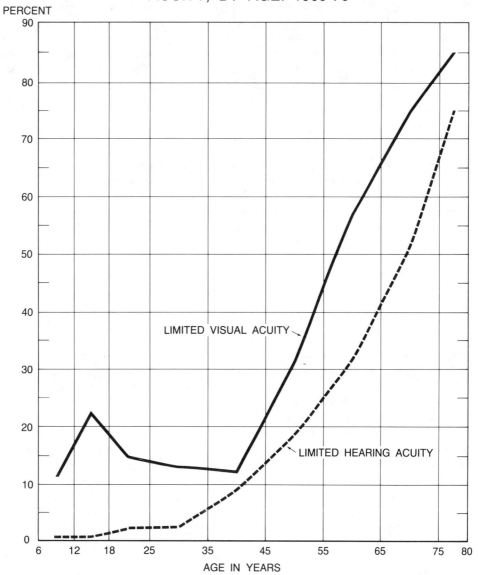

Figure 4-17. A corrected version of Figure 4-16. It will be observed that there are four different sizes of age categories—five 10-year intervals, two 6-year, one 7-year, and one 4-year interval. In order to emphasize these differences, the data have been plotted at the midpoint of each interval.

uals or families with wages or income below stated levels. Figure 4-14 indicates the percentage of families with incomes below $1000, below $2000, and so on. An important advantage of the cumulative frequency curve is its special utility and reliability, regardless of the size of class intervals. In Figure 4-14, class intervals of $1000 were used for part of the curve and larger ones for other parts. (See Table 4-1.)

LORENZ CURVE

The Lorenz curve[11] is at least a distant relative of the ogive. Both are concerned with the distribution of frequency series. In constructing an ogive the accumulated frequencies of a single distribution, expressed either in percentages or in whole numbers, are plotted either from the lowest interval ("less than" ogive) or from the highest ("more than" ogive). In a Lorenz curve there are two frequency distributions, one plotted on the X-axis and one on the Y-axis, each as percentages of the total. It will be noted that Figure 4-15 is based on the two distributions in Table 4-1. Each pair of percentages is plotted on the chart and are indicated by a dot. For example, the Lorenz curve and the figures in Table 4-1 show that the "poorest" 20.0 percent of the families had personal incomes of less than $3000, representing 6.4 percent of total income.

A careful examination of Figure 4-15 reveals that the Lorenz curve can be read upward, downward, or sidewise. Reading upward, for example, indicates that the lowest 40 percent of the families received appoximately 19 percent of the income; reading downward, the top 10 percent of the families received about 30 percent of the income; reading from left to right, the "poorest" 41 percent of the families, 20 percent of the income; and from right to left, the "richest" 5 percent of the families, 20 percent of the income.[12] The dashed diagonal line on the chart—the "line of equal distribution"—indicates equality of income for all families. The area between the dashed line and the curve is a measure of income inequality.

LINE CHARTS WITH UNEQUAL CLASS INTERVALS

One of the most common errors in designing line charts such as frequency distributions and series of rates with unequal class intervals is the confusion that often occurs in designating and spacing scale units for the X-axis. Figure 4-16 is an example of a chart with

class intervals that are improperly spaced. It will be seen that the spacing between every age designation is identical, although there are four different sizes of class intervals—5, 6, 7, and 10 years. Figure 4-17 was redesigned correctly in order to show these differences. Moreover, to emphasize the different sizes of class intervals, the data in Figure 4-17 have been plotted at the midpoint of each class interval. Figure 4-18 is another example in which the class-interval divisions are grossly in error. The spacing for all of the intervals are equal although there are two that represent 1 week; one, 8 weeks; one, 4 weeks; one, 3 weeks and one, open-ended. Also, the mechanical features of the chart, particularly the curves and the lettering, would be classed as inferior.

SCATTER DIAGRAMS AND CORRELATION MATRICES

In designing a scatter diagram (also called scattergram or scatter plot), it is not an uncommon practice to construct the basic pattern in the form of a rectangle rather than a square. In comparison to most design errors and deficiencies occurring in statistical graphics, this practice probably would be considered of minor significance. However, it is apparent that a square is preferable to a rectangle in the basic design pattern of a scatter diagram, since it provides greater clarity to the relationship between variables, as well as facilitating interpretation. Also, it can be said that the greater the degree of elongation of the rectangle, the less desirable the scatter diagram becomes as a medium of visual communication.

Figure 4-19 is an example of a scatter diagram whose basic pattern is rectangular in shape. It shows the relationship between income and the proportion of starchy foods (grain products, roots, and tubers) in the diets for 27 countries throughout the world. It is clear from this chart that there is a marked negative relationship between the proportion of starchy foods in the diet and income. That is, the greater the income, the less the proportion of starchy foods in the diet.

---

[11] Named after M. O. Lorenz who invented this curve. See the original paper entitled "Methods of Measuring the Concentration of Wealth," *Journal of the American Statistical Association,* New Series No. 70 (June 1905), 209–219.

[12] Frederick V. Waugh, *Graphic Analysis in Agricultural Economics,* Washington, D.C.: U.S. Department of Agriculture, Agricultural Marketing Service, 1957, p. 8.

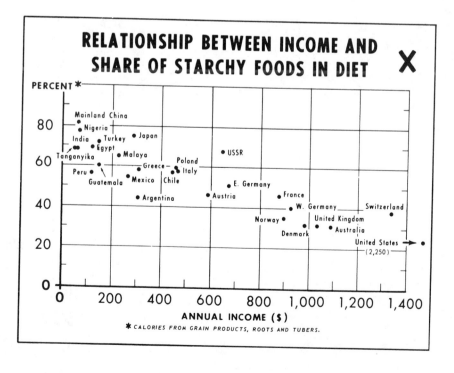

Figure 4-18. *Another example of a poorly designed chart with class intervals of six different sizes. However, all of the spacings on the X-axis are the same: In addition to the erroneous spacing for the class intervals, the drafting quality of the chart is clearly substandard. [From Factors Associated with Low Birth Weight: United States, 1976, DHEW Publication No. (PH5) 80-1915, Hyattsville, Md.: National Center for Health Statistics, 1980, p. 12.]*

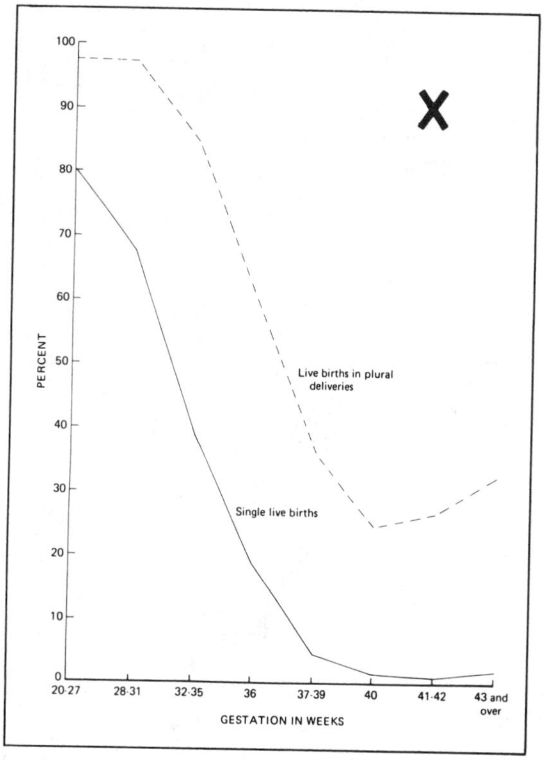

Percent of infants of low birth weight among single live births and among live births in plural deliveries by period of gestation: Total of 42 reporting States and the District of Columbia, 1976

Figure 4-19. *An example of a scatter diagram in rectangular form. Except for the basic rectangular pattern, this chart would rate relatively high in quality. (From Lester R. Brown, Man, Land and Food, Foreign Agricultural Economic Report No. 11, Washington, D.C.: U.S. Department of Agriculture, 1963, p. 45.)*

INTERCORRELATION OF INDICES OF VOTING BEHAVIOR, SEATTLE: 1930 TO 1940

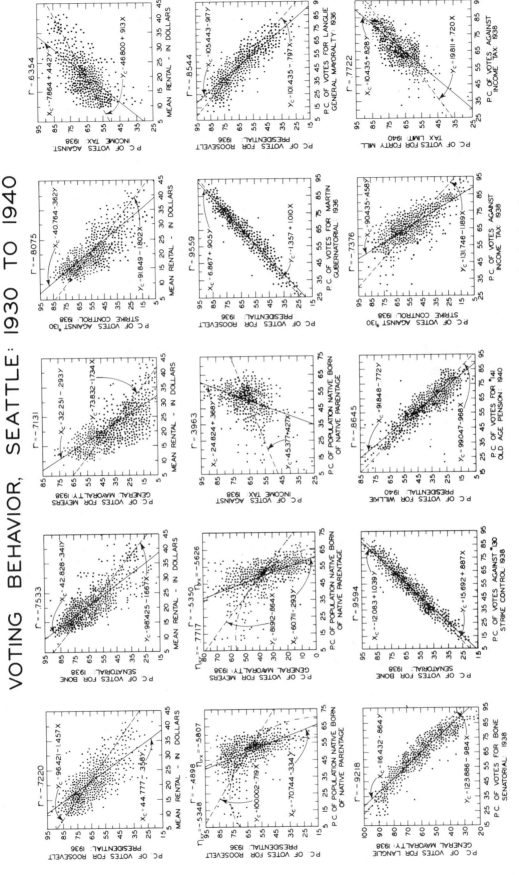

Figure 4-20. A series of 15 scatter diagrams portraying voting behavior in a large urban community. The data for each distribution are based on over 500 election precincts. The basic pattern of each of the 15 scatter diagrams is a square. Also, note the coefficients of correlation and regression lines. (From Calvin F. Schmid, Social Trends in Seattle, Seattle: University of Washington Press, 1944, p. 274.)

INTERCORRELATIONS OF TEMPORAL PATTERNS OF CRIME RATES
THREE CITIES OF WASHINGTON AND OTHER AREAS: 1940-1969

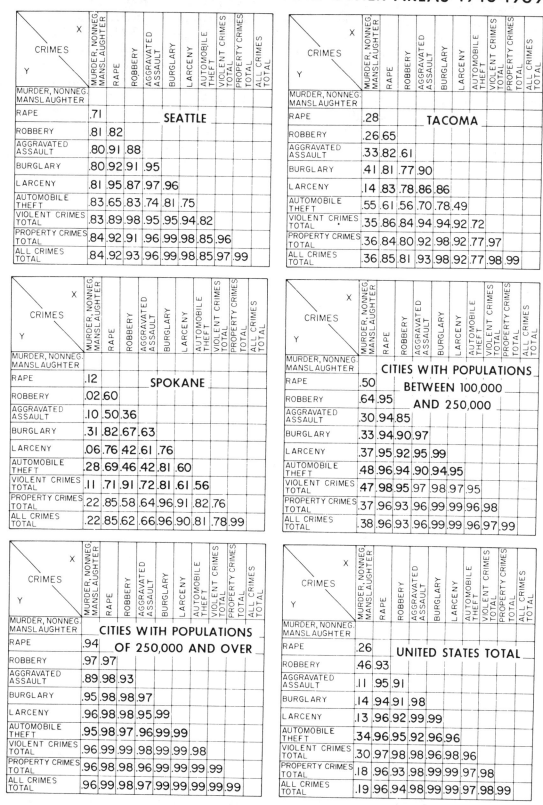

Figure 4-21. *Six correlation matrices that portray interrelations of temporal trends and patterns of various crimes for different areas and groupings. Although the correlation matrix has many of the characteristics of a statistical table it can be designed to serve as a vehicle of visual communication. (From Calvin F. Schmid and Stanton E. Schmid, Crime in the State of Washington, Olympia: Law and Justice Planning Office, Planning and Community Affairs Agency, 1972, p. 23.)*

Unlike Figure 4-19, which includes only 27 cases, Figure 4-20 includes over 500 cases (election precincts) in each of 15 scatter diagrams. Moreover, the basic structure of the 15 scatter diagrams is a square with the inclusion of regression lines and coefficients of correlation. In two instances, curvilinear regression lines and coefficients of correlation were computed also. These scatter diagrams are part of a fairly extensive analysis of voting patterns as they relate to economic, demographic, and social factors in a large urban community.

Although a correlation matrix generally exhibits more of the characteristics of a statistical table than of a statistical chart, frequently it can be designed to serve as an important tool of visual communication. With careful planning, a correlatrion matrix can be constructed to exemplify a substantial number of basic graphic characteristics and qualities of an effective statistical chart. For example, Figure 4-21 was designed in part to summarize a discussion of a 30-year trend in crime for six different areas. These matrices summarize fairly graphically the similarities, differences, and interrelations of trends and patterns of major crimes for six different areas and groupings.

THE SEMILOGARITHMIC CHART IS ONE OF the most useful, reliable, and widely adaptable of all statistical charts. First, it is unequaled for comparing proportional rates of change for a number of different curves or for several segments of the same curve. In contrast, the arithmetic chart compares absolute differences or increments of change. The semilogarithmic chart emphasizes relative or percentage change. At the same time, the calibrations and scale figures on the vertical scale provide a ready reference to the actual values that have been plotted. Second, the semilogarithmic chart represents a means for comparing with clarity and reliability two or more series that differ widely in absolute values. Regardless of the amounts indicated on the scale, the slopes manifested by the curves correctly indicate rates of change. Third, the semilogarithmic chart provides a sound technique for comparing several series of data that are not expressed in common units. This can be accomplished very simply and without distortion or deception. These and other characteristics of the semilogarithmic chart will be discussed in more detail in the following pages.

Paradoxically, there seems to be a pervasive impression that the probability of malpractice and misinterpretation is greater for semilogarithmic charts than for most other graphic forms. Perhaps the unusual label it bears or its distinctive logarithmic scale contribute to its enigmatic and even formidable status. Every proficient chartmaker, of course, should be thoroughly conversant with the semilogarithmic chart, including its advantages, limitations, and applications. Also, in viewing the total communication process it is essential for the chartmaker to have some assurance that the user possesses at least a superficial understanding of semilogarithmic charts and can interpret them correctly. The cartographer, for example, has observed repeatedly from experimental and other evidence that the communication process can break down completely as a consequence of the ineptitude of either the chartmaker or the user.

CHARACTERISTICS OF THE SEMILOGARITHMIC CHART

It will be recalled from the chapter on rectilinear coordinate charts that equal distances along an arithmetic scale represent equal amounts or increments of change. However, on a logarithmic scale, equal distances represent equal rates of change. For example, an increase from 250 to 300 on an arithmetic scale

CHAPTER FIVE

SEMILOGARITHMIC CHARTS

Their Use and Misuse

would show an increase in amount of 10 times as great as an increase of 25 to 30, but on a semilogarithmic chart the two increases, as gauged by the slope of the curves, would be exactly equal since their respective rates of change are identical. Accordingly, the emphasis in a semilogarithmic chart is on rates or ratios of change. Incidentally, because of this fact, semilogarithmic charts are frequently referred to as ratio charts.

Graphically, logarithms can be used to portray rates of change in two different ways: first, by plotting the logarithms of a series of numbers on a natural scale, and second, by plotting the actual numbers in a series on a logarithmic scale. The second procedure is far more practical and convenient, and it is the one that is generally used.[1] Various kinds of commercially printed semilogarithmic-ruled paper are readily available from most stationery and art supply retail outlets.

Perhaps the most distinctive structural characteristic of the semilogarithmic chart lies in its vertical scale, which is generally logarithmic. The horizontal scale is arithmetic, which means that the intervals are equally spaced.[2]

---

[1] For a discussion pertaining to the construction of semilogarithmic charts, see Calvin F. Schmid and Stanton E. Schmid, *Handbook of Graphic Presentation*, New York: John Wiley & Sons, 1979, pp. 101–106.

[2] On rare occasions, the axes may be reversed, particularly in plotting frequency distributions. See Calvin F. Schmid and Stanton E. Schmid, *Handbook of Graphic Presentation*, New York: John Wiley & Sons, 1979, pp. 133–134.

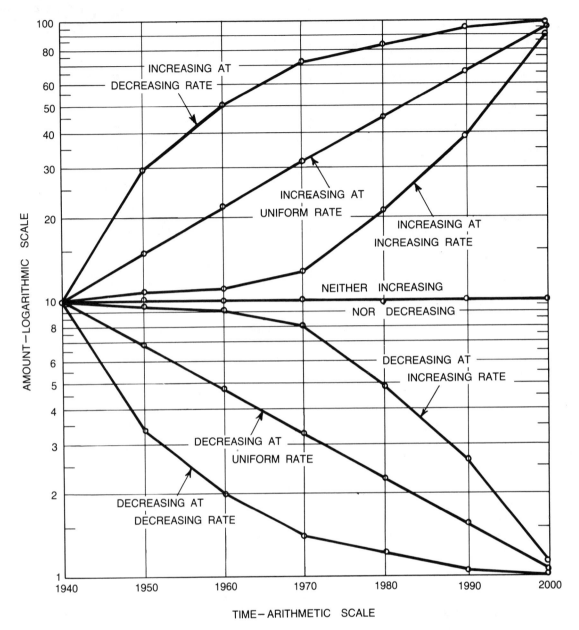

Figure 5-1. Interpretation of typical curve patterns for the semilogarithmic chart. See text for additional comments.

Semilogarithmic rulings are composed of one or more "cycles," "phases," "tiers," "banks," or "decks." One cycle covers the range of values represented by only one power of ten. Accordingly, for example, one cycle could be represented by 10^1 to 10^2, or from 10 to 100; two cycles conforming to the preceding sequence would accommodate values from 10^1 to 10^3, or from 10 to 1000; three cycles, from 10^1 to 10^4, or from 10 to 10,000; and four cycles, from 10^1 to 10^5, or from 10 to 100,000. Theoretically, the number of cycles can extend indefinitely in either direction, and the ranges are always expressed as multiples or divisions of 10.

In actual practice, the number of cycles in a chart is determined by the values of the highest and lowest items of the series that are to be plotted. If the range of values in a series is relatively narrow, it is proper to use only part of a cycle. For example, if the range of values to be plotted is from 220 to 574, then only the portion of a cycle from 200 to 600 would be constructed. In such a case it would not be necessary to indicate that the cycle is incomplete by a broken line or by some other special marking such as would be required in the case of an incomplete arithmetic scale. In a logarithmic scale there is no zero base line, and the interpretation of curves in semilogarithmic charts is not conditioned by distance from a base line as typified in arithmetic charts. Also, since there is no zero on a semilogarithmic charts, it is not possible to pass from positive to negative numbers or vice versa.

INTERPRETATION OF SEMILOGARITHMIC CHARTS

The key principle in interpreting a semilogarithmic chart is based on the fact that the relative steepness or slope of a curve, whether ascending or descending, indicates the rate of change of a variable. This principle is applicable to the general trend of a curve as a whole as well as to segments or portions of a curve. The steeper the slope of a curve, either ascending or descending, the more pronounced the rate of change.

The interpretation of typical semilogarithmic curve patterns is summarized in Figure 5-1 as well as in the following explanatory statements:

1. A curve that is horizontal is neither increasing or decreasing.

2. A positive or upward slope indicates that the variable is increasing, and a negative or downward slope, decreasing.

3. A curve that is ascending or descending in a straight line is changing at a constant rate. A series of this kind represents a geometric progression.

4. As indicated in the preceding paragraph, the steeper the slope of a curve, the greater the rate of change.

5. Two or more curves with the same distance between them (parallel), although of different magnitude, show the same rate of change.

6. If a variable is increasing (ascending) at an increasing rate, the general configuration of the curve will be concave to the horizontal axis.

7. If a variable is increasing (ascending) at a de-

Figure 5-2.   A special protractor devised to measure percentage increase or decrease in curves in a semilogarithmic chart. (From Australian Bureau of Statistics, Seasonally Adjusted Indicators, **Canberra, Australia, 1976, p. 11.**

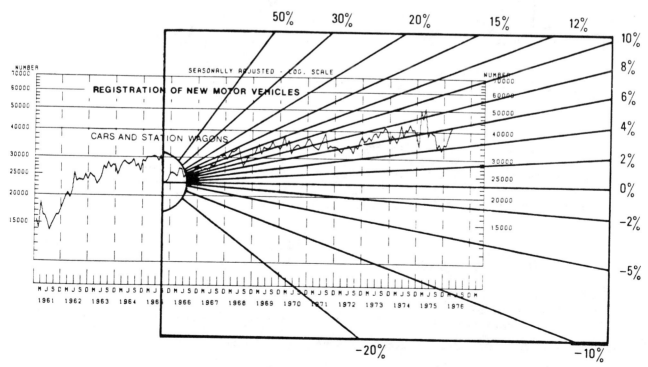

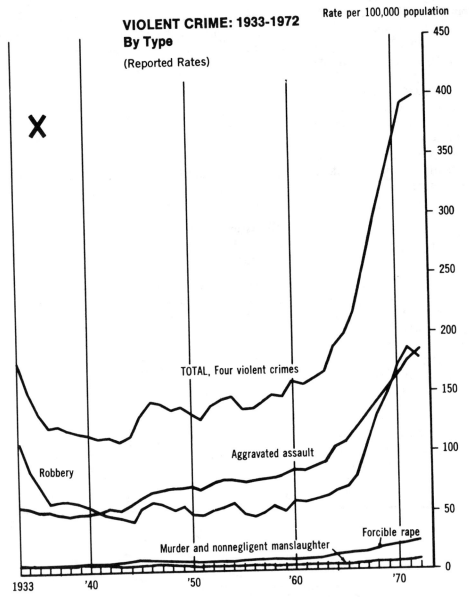

VIOLENT CRIME: 1933-1972
By Type

(Reported Rates)

Rate per 100,000 population

Figure 5-3. An example of a graphic form inappropriate for presenting several temporal series that differ widely in values. Because of their low rates, the curves for murder and nonnegligent manslaughter and for rape possess virtually no significance. In fact, as a technique for showing trends of this kind, the entire chart is most unsatisfactory. (From Executive Office of the President, Office of Management and Budget, Social Indicators, 1973, Washington, D.C.: Government Printing Office, 1973, p. 44.)

creasing rate, the general configuration of the curve will be convex to the horizontal axis.

8. If a variable is decreasing (descending) at an increasing rate, the general configuration of the curve will be concave to the horizontal axis.

9. If a variable is decreasing (descending) at a

decreasing rate, the configuration of the curve will be convex to the horizontal axis.

---

[3] Calvin F. Schmid and Stanton E. Schmid, *Handbook of Graphic Presentation*, New York: John Wiley & Sons, 1979, pp. 113–115.

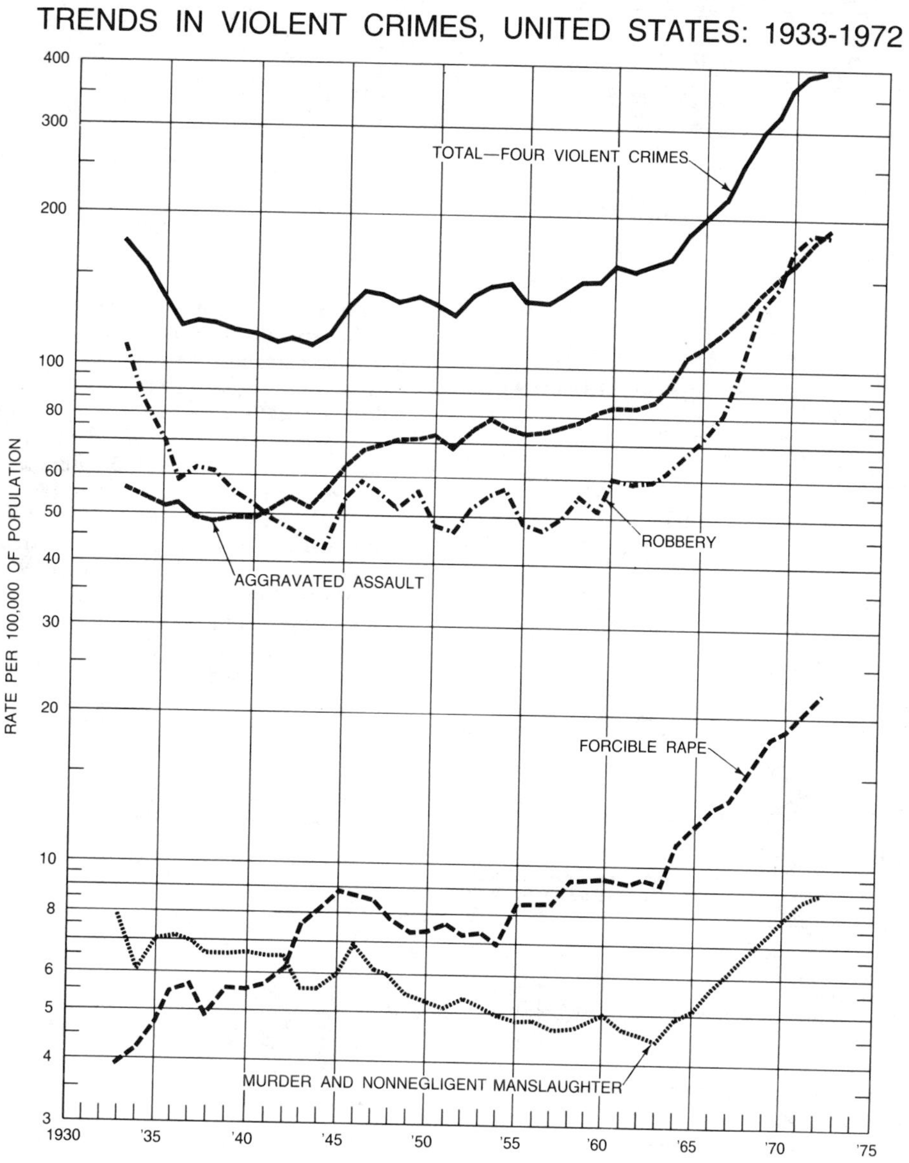

TRENDS IN VIOLENT CRIMES, UNITED STATES: 1933-1972

Figure 5-4. This chart is a redesigned semilogarithmic version of Figure 5-3. The semilogarithmic chart clearly demonstrates that the visual message conveyed by the five series of crime data can be meaningful, unambiguous, and reliable.

TECHNIQUES FOR DETERMINING PERCENTAGE CHANGE IN SEMILOGARITHMIC CHARTS

Since semilogarithmic charts emphasize rates of change, certain simple mechanical techniques have been devised to estimate percentage change with reasonable close approximation. One method may be found in our *Handbook of Graphic Presentation.*[3]

Another less precise, but perhaps more flexible method is illustrated by Figure 5-2. It will be observed that it is in the form of a simple protractor designed specifically for a particular chart or series of

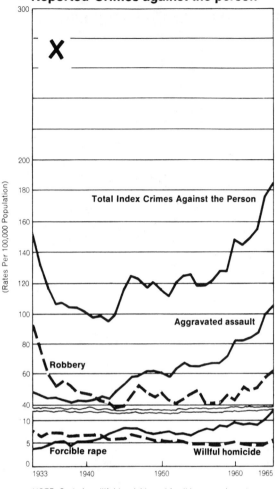

Index Crime Trends, 1933-1965
Reported Crimes against the person

(Rates Per 100,000 Population)

Total Index Crimes Against the Person

Aggravated assault

Robbery

Forcible rape   Willful homicide

1933 1940 1950 1960 1965

NOTE: Scale for willful homicide and forcible rape enlarged,
to show trend.
Source: FBI, Uniform Crime Reports Section; unpublished data.

Figure 5-5. A clumsy attempt to adjust for the serious limitations of an arithmetic line chart designed to portray several series of data representing wide ranges in value. Actually this chart is no improvement over Figure 5-3, which has a single ordinal scale. In Figure 5-5, there are two vertical scales, but they are drawn incorrectly as a single broken scale with entirely different scale intervals. Also, the upper one third of the grid is superfluous. (From the President's Commission on Law Enforcement and Administration of Justice, Task Force Report: Crime and its Impact—An Assessment, Washington, D.C.: Government Printing Office, 1967. p. 19.)

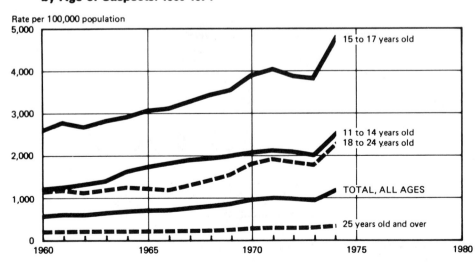

Urban Police Arrests of Suspected Offenders for Crimes of Theft, by Age of Suspects: 1960-1974

Rate per 100,000 population

15 to 17 years old

11 to 14 years old
18 to 24 years old

TOTAL, ALL AGES

25 years old and over

1960 1965 1970 1975 1980

Figure 5-6. Another example of an arithmetic chart that has been chosen to portray several series of data that show relatively wide variation in values. The implications of using an arithmetic chart for this purpose can be understood more clearly by comparing this figure with Figure 5-7. On the basis of such a comparison, it is obvious that the semilogarithmic chart is much more effective. (From U.S. Department of Commerce, Office of Federal Statistical Policy and Standards, Social Indicators, 1976, Washington, D.C.: Government Printing Office, 1977, p. 234.)

charts with identical scales. The protractor is drawn on transparent plastic and can be superimposed readily on any time period of the chart. The zero point of the protractor is placed over the chosen starting point on the curve with its vertical line parallel to the vertical lines on the chart. A reading of percentage change is then taken at the point on the curve at which the observation is intended to end.[4]

SPECIAL APPLICATIONS AND ADVANTAGES OF THE SEMILOGARITHMIC CHART

An examination of the four succeeding charts clearly indicates the demonstrable superiority of the semi-

logarithmic chart in comparison to the arithmetic line chart in portraying multiple series of temporal data that vary widely in values. The values of the series included in Figures 5-3 and 5-4 range from 3.9 to 397.7 per 100,000 of population, and in Figures 5-6 and 5-7, from 206.9 to 4,734.9 per 100,000 of population. For example, even in terms of the roughest approximation, it is impossible to derive any meaningful conception of what the values and trends are for homicide and rape by examining the arithmetic chart (Figure 5-3). At most, one can justifiably conclude that the rates for these two crimes are low in com-

---

[4] Australian Bureau of Statistics, *Seasonally Adjusted Indicators,* Canberra, Australia, 1976, p. 11.

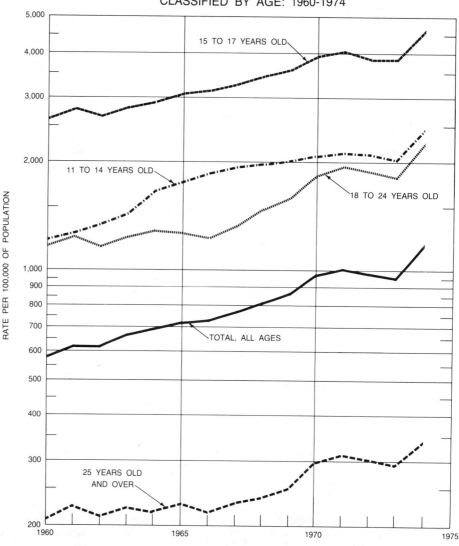

URBAN POLICE ARRESTEES FOR THEFT, CLASSIFIED BY AGE: 1960-1974

Figure 5-7. A reconstruction of Figure 5-6 as a semilogarithmic chart. It is apparent that for depicting graphically the data pertaining to trends in age differentials of criminal behavior, the semilogarithmic chart is superior to the arithmetic chart.

parison to the other crime categories shown on the chart, which, of course, possess only the slightest significance in a presentation of this kind. It is obvious that the arithmetic chart is an ill-chosen graphic form for presenting data of this kind. As a medium of visual communication for these data it is a complete failure. By contrast, the semilogarithmic chart portrays rates of change for all of the five crime categories in a clear, reliable, and comparable manner. There are no ambiguities, uncertainties, or distortions. The portrayal of the relatively low rates for murder and nonnegligent manslaughter and forcible rape are just as authentic and meaningful in the semilogarithmic chart as the much higher rates shown by the three other categories.[5]

With the same series of data as Figure 5-3, it is interesting to find in another report that an effort was made to compensate for the deficiencies of the arithmetic chart (Figure 5-5). Unfortunately, the adjustments that were made in Figure 5-5 are erroneous and misleading, with consequences worse than the original chart. It will be seen that the vertical axis has been divided into two different and incompatible scales, but the manner in which they have been drawn gives the false impression of identity and continuity. However, below the break, the scale intervals represent units of 5 and above the break, units of 20. The spacing of the units of the upper portion is only slightly larger than the spacing of units in the lower portion. In addition, over a third of the upper portion of the chart is blank and superfluous.

Figures 5-6 and 5-7 reaffirm what was said about the two preceding charts, particularly the authenticity and meaningfulness of the semilogarithmic chart for displaying several series of temporal data that manifest a wide range of values. Both the arithmetic and semilogarithmic charts indicate a close as well as consistent relationship between criminal behavior and age. The age group 15–17 years exhibits by far the highest annual rates for the entire 14-year period, while the age group 25 years and over ranks lowest. However, the age differential rates as well as their consistent patterns are revealed much more clearly and correctly by the semilogarithmic chart than by the arithmetic chart.

Figure 5-8 is another example of the flexibility, versatility, and adaptability of the semilogarithmic chart. It will be observed that there are nine curves on this chart with values ranging from less than 60 to almost 50,000. Without fear of contradiction, it can be said that there is no other way to portray graphically so much data so succinctly and in such a complete, reliable, and readily interpretable manner. By comparison, an attempt to display these data on an arithmetic grid would result in meaningless confusion. Figure 5-8 carries a heavy load of data without introducing clutter or distracting complexities. Judged in terms of graphic design standards including scales, scale designations, scale lines, scale numerals, curve patterns, title, curve legends, balance, harmony and proportions, the chart would rate relatively high.[6]

Figure 5-9 also indicates the versatility of the semilogarithmic chart by presenting simply and logically a 70-year capsuled historical analysis of college and university enrollment trends in relation to the major college-age population pool in the state of Washington. Because of their relevance and importance in the interpretation of trends and relationships, both population and enrollment figures have been superimposed on the chart, thus facilitating both relative and absolute comparisons. A clearly differentiated 15-year forecast period for both population and enrollment is also shown on the chart. The relative slope of the two curves, so characteristically significant in interpreting semilogarithmic charts, graphically shows the rapid rate of growth of college and university enrollment in relation to college-age population.

Figure 5-10 is illustrative of another useful and significant application of the semilogarithmic chart for displaying comparative trends of two or more series of variables that are intrinsically different, for example, educational status of the population, income, quality of dwelling units, and crime rates. Sometimes a multiple-scale arithmetic chart may be used to show temporal series of more than one variable, but very frequently with a risk of distortion or even complete failure, especially if there are more than two variables. In portraying data of this kind, the semilogarithmic chart is unquestionably superior to the arithmetic chart.

In Figure 5-10, there are three series of data: pop-

---

[5] It is recognized, of course, that Figure 5-3 is poorly designed as well as being deficient in other respects. Nothing has been stated concerning this fact since the central problem under discussion in this section pertains to the improper choice and misapplication of arithmetic charts rather than their design and quality per se. The problems of design efficiency and quality of arithmetic charts are covered in considerable detail in Chapter 2.

[6] The terminology and abbreviations used in this chart are fully explained in the text of the monograph in which this chart appeared. For example, U. of W. means University of Washington; W.W.S.C., Western Washington State College.

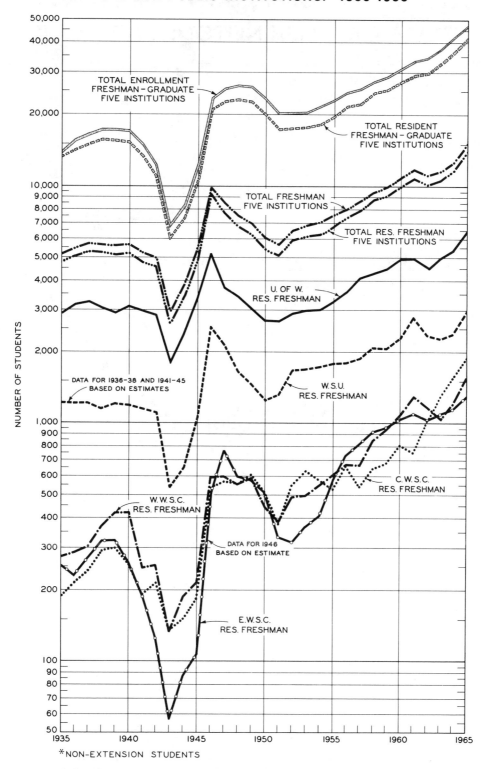

TRENDS IN RESIDENT FRESHMAN ENROLLMENT*
FIVE MAJOR PUBLIC INSTITUTIONS: 1935-1965

NUMBER OF STUDENTS

TOTAL ENROLLMENT
FRESHMAN – GRADUATE
FIVE INSTITUTIONS

TOTAL RESIDENT
FRESHMAN – GRADUATE
FIVE INSTITUTIONS

TOTAL FRESHMAN
FIVE INSTITUTIONS

TOTAL RES. FRESHMAN
FIVE INSTITUTIONS

U. OF W.
RES. FRESHMAN

DATA FOR 1936-38 AND 1941-45
BASED ON ESTIMATES

W.S.U.
RES. FRESHMAN

W.W.S.C.
RES. FRESHMAN

C.W.S.C.
RES. FRESHMAN

DATA FOR 1946
BASED ON ESTIMATE

E.W.S.C.
RES. FRESHMAN

*NON-EXTENSION STUDENTS

Figure 5-8. A practical example of the flexibility and versatility of the semilogarithmic chart. This chart portrays nine detailed series of data with accuracy and lucidity. (From Calvin F. Schmid, Vincent A. Miller, and William S. Packard, Enrollment Statistics, Colleges and Universities, State of Washington, Fall Term, 1965, Seattle: Washington State Census Board, 1966, p. 76.)

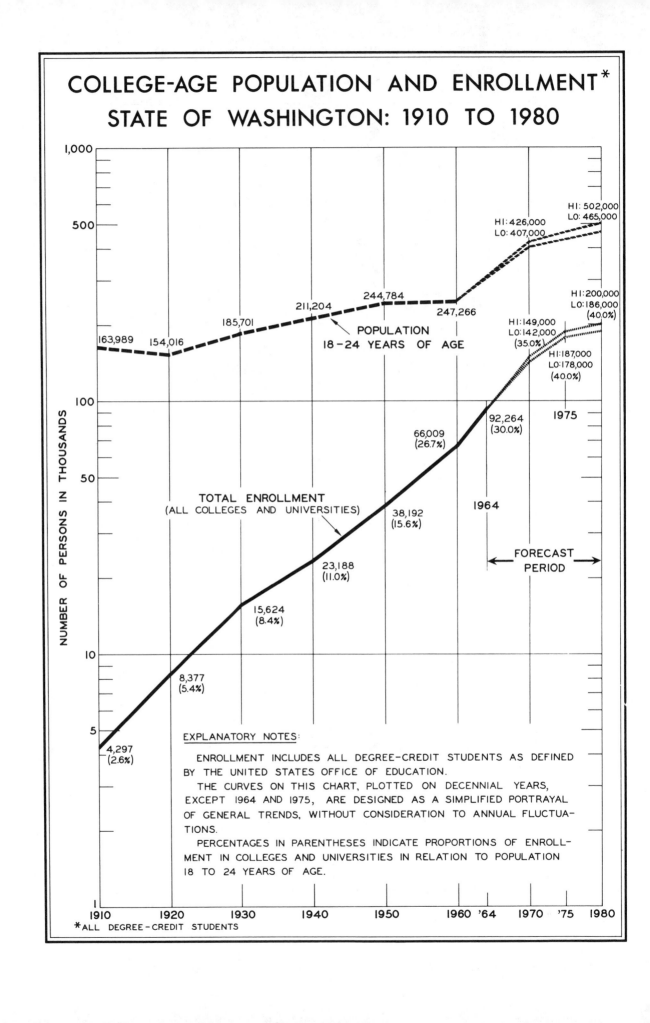

COLLEGE-AGE POPULATION AND ENROLLMENT*
STATE OF WASHINGTON: 1910 TO 1980

NUMBER OF PERSONS IN THOUSANDS

POPULATION
18-24 YEARS OF AGE

163,989   154,016   185,701   211,204   244,784   247,266

HI: 426,000
LO: 407,000

HI: 502,000
LO: 465,000

HI: 200,000
LO: 186,000
(40.0%)

HI: 149,000
LO: 142,000
(35.0%)

HI: 187,000
LO: 178,000
(40.0%)

TOTAL ENROLLMENT
(ALL COLLEGES AND UNIVERSITIES)

66,009
(26.7%)

92,264
(30.0%)

1975

1964

38,192
(15.6%)

23,188
(11.0%)

15,624
(8.4%)

FORECAST
PERIOD

8,377
(5.4%)

4,297
(2.6%)

EXPLANATORY NOTES:

ENROLLMENT INCLUDES ALL DEGREE-CREDIT STUDENTS AS DEFINED
BY THE UNITED STATES OFFICE OF EDUCATION.
THE CURVES ON THIS CHART, PLOTTED ON DECENNIAL YEARS,
EXCEPT 1964 AND 1975, ARE DESIGNED AS A SIMPLIFIED PORTRAYAL
OF GENERAL TRENDS, WITHOUT CONSIDERATION TO ANNUAL FLUCTUA-
TIONS.
PERCENTAGES IN PARENTHESES INDICATE PROPORTIONS OF ENROLL-
MENT IN COLLEGES AND UNIVERSITIES IN RELATION TO POPULATION
18 TO 24 YEARS OF AGE.

1910   1920   1930   1940   1950   1960 '64   1970   '75   1980

*ALL DEGREE-CREDIT STUDENTS

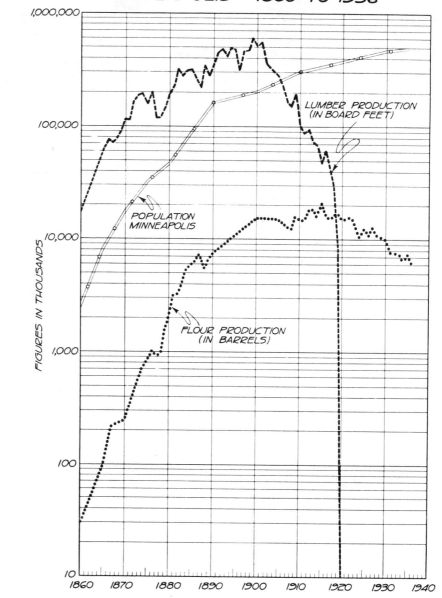

GROWTH AND DECLINE OF
LUMBERING AND FLOUR MILLING
MINNEAPOLIS : 1860 TO 1936

Figure 5-10. This chart presents three series of data that are expressed in entirely different units. One series represents population (persons); another, lumber production (board feet); and the third, flour production (barrels). (*From Calvin F. Schmid, Social Saga of Two Cities, Minneapolis: Minneapolis Council of Social Agencies, 1937, p. 15.*)

Figure 5-9. Another example of the wide application and utility of the semilogarithmic chart. Its simplicity, clarity, and reliability combine to make it an extraordinarily effective medium of visual communication. (*From Calvin F. Schmid et al., Enrollment Forecasts, State of Washington: 1965 to 1985, Seattle and Olympia: Washington State Census Board and Washington State Department of Commerce and Economic Development, 1966, p. 10.*)

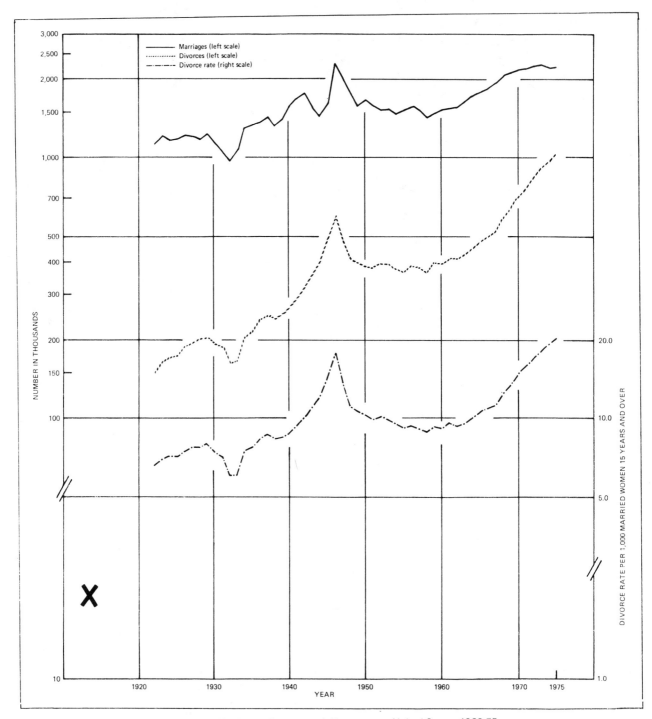

Marriages, divorces, and divorce rates: United States, 1922-75

Figure 5-11. *An example of a semilogarithmic chart that manifests certain short-comings. A redesigned version is presented by Figure 5-12. See text for additional comments.* [*From Alexander A. Plateris,* Divorces by Marriage Cohort, *DHEW Publication No. (PHS) 79-1912, Hyattsville, Md.: National Center for Health Statistics, 1979, p. 4.*]

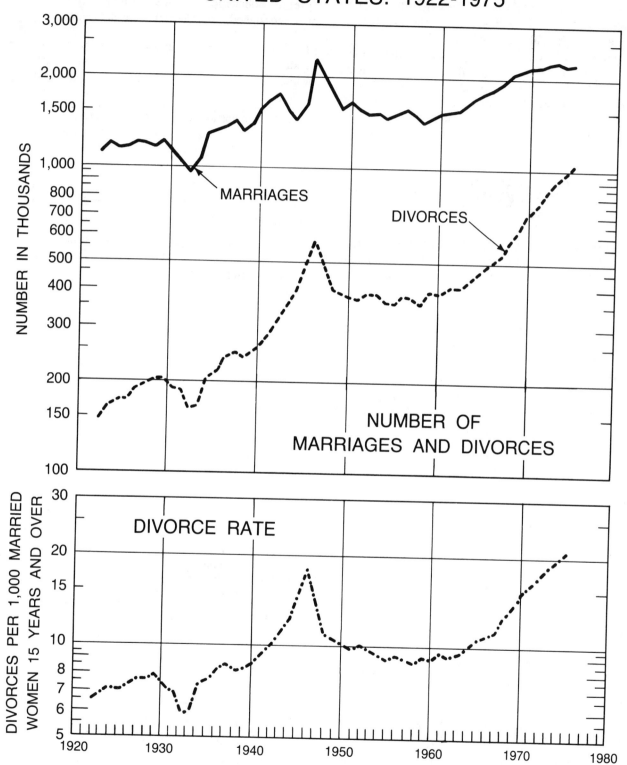

MARRIAGES, DIVORCES, AND DIVORCE RATES
UNITED STATES: 1922-1975

NUMBER IN THOUSANDS

3,000

2,000

1,500

1,000

800
700
600
500

400

300

200

150

100

MARRIAGES

DIVORCES

NUMBER OF
MARRIAGES AND DIVORCES

DIVORCES PER 1,000 MARRIED WOMEN 15 YEARS AND OVER

30

20

15

10

8
7
6
5

DIVORCE RATE

1920   1930   1940   1950   1960   1970   1980

Figure 5-12.   *This chart is a reconstruction of Figure 5-11. See text for a discussion of the revisions that have been made.*

ulation (number of persons), lumber production (number of board feet), and flour production (number of barrels). More curves, of course, could have been added if this had been considered desirable or relevant for a fuller elucidation of the problem at hand. The vertical scale consists of five cycles, ranging in value from 10 to 1,000,000. Each series of data is simply plotted on the grid in accordance with its particular values, thus providing an accurate and readily interpretable comparison of growth and decline of the lumber and milling industries as well as the growth of population in the city of Minneapolis.

EXAMPLES OF DESIGN PROBLEMS AND SUGGESTED SOLUTIONS

Figures 5-11 and 5-12 present the same series of data: number of marriages, number of divorces, and the divorce rate per 100 married women 15 years of age and over for the United States from 1922 to 1975.

Figure 5-11 exhibits a number of design shortcomings, whereas Figure 5-12 is a revised version embodying suggested corrections and improvements. In order to understand more clearly why Figure 5-11 has been redesigned, note the following deficiencies:

1. There is a superfluous amount of space especially on the left-hand lower portions of the chart, which detracts from its overall composition and balance.

2. The two vertical scales are indicated with slash marks as being broken. This practice in both instances is not only unconventional, but unnecessary.

3. Since the plotted data have been compiled on an annual basis, the fact should be indicated on the horizontal axis by tick marks and scale lines.

4. The curves, perhaps the most important elements of the chart, are shown in thin lines, thus giving the impression of weakness and unimportance.

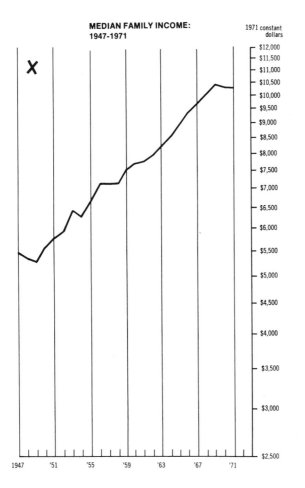

Figure 5-13.   Another case of poor design. See text for a list of deficiencies. (From the Executive Office of the President, Office of Management and Budget, Social Indicators, 1973, Washington, D.C.: Government Printing Office, 1973, p. 152.)

5. The lettering for the curve designations is too small in light of the purpose it serves. Furthermore, the inconspicuous location of the curve designations is a negative factor in the total chart design.

6. Also, there should be more scale lines and ticks to facilitate easier and more accurate interpretation of the logarithmic values.

7. In general, the lettering is too small and out of proportion to the size of the chart.

8. Again, in general, the drafting quality of the chart reflects mediocrity.

Figures 5-13 and 5-14 are two semilogarithmic charts: One appears as it was originally published in *Social*

Indicators 1973 (Figure 5-13) and the other is a redesigned version (Figure 5-14). It will be observed that in redesigning Figure 5-13 the following changes have been made:

1. Elimination of a large portion of the grid that was superfluous. The lowest vertical scale value is now $500 rather than $2500. Also, a small unused portion of the upper part of the chart has been reduced in size. The range of the data in the series is from $5278 to $10,423.

2. The major scale divisions on the horizontal axis were changed to five-year intervals, beginning with 1945 and with scale-points for single years. Basic divisions reflecting equal quinquennial intervals are more conventional, convenient, and in-

MEDIAN FAMILY INCOME, UNITED STATES: 1947-71

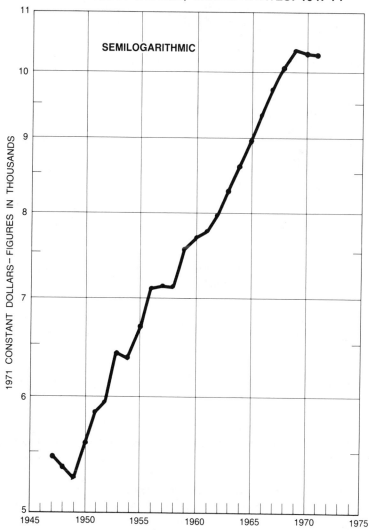

Figure 5-14. This chart is a revision of Figure 5-13. In Figure 5-14, almost one-half of the grid shown in Figure 5-13 has been eliminated, and as a consequence the vertical scale (logarithmic) extends only from $5000 to $11,000. The full range of the data runs from $5278 to $10,423. Since the spacing for the vertical scale has been increased substantially, the slope of the curve in Figure 5-14 is much steeper than in Figure 5-13.

MEDIAN FAMILY INCOME, UNITED STATES: 1947-71

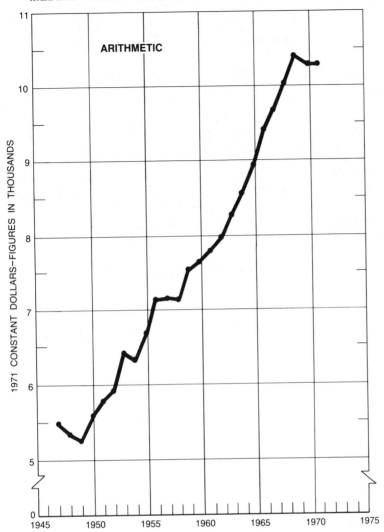

Figure 5-15. This chart, a rectilinear coordinate line chart, portrays the same data as in Figures 5-13 and 5-14. In order to make valid comparisons with Figure 5-14, the overall size and proportions of Figure 5-15 were constructed the same as in Figure 5-14. It is readily observable that the configurations and slopes of the curves in these charts are markedly similar. Such similarities may occur when the data show a relatively narrow range as well as regular and consistent trends. In instances of this kind, the designer must decide which is preferable for the purpose at hand: an arithmetic or a semilogarithmic chart.

terpretable than odd numbered years, which are indicated on the original chart.

3. The scale figures for the vertical axis have been moved to the left side of the grid. Some of the scale figures have been eliminated.

4. Scale lines have been added to reveal the values more clearly as well as to facilitate interpretation of the chart.

Although the utility and versatility of the semilogarithmic chart have been emphasized in this chapter, it should not be assumed that the semilogarithmic chart can serve as a complete replacement for the arithmetic chart. Each type of chart has very different characteristics and applications and the choice of one or the other should be gauged mainly within the con-

text of the problems and objectives at hand. Perhaps, strange as it may seem, there are rare instances where the selection of either a semilogarithmic or an arithmetic chart would make little difference since they would exhibit a pronounced similarity. So far as the planning of *Social Indicators 1973* is concerned, it would be of some interest to ascertain why a semilogarithmic chart was chosen for Figure 5-13, especially since so many arithmetic charts throughout the book were selected in preference to semilogarithmic charts, which in a large number of instances would have been more appropriate. After a further examination of the data on median family income, it was hypothesized that curves plotted on either an arithmetic or semilogarithmic grid would not be radically different. Accordingly, in order to make a valid comparison, data were plotted on an arithmetic grid (Figure 5-15).

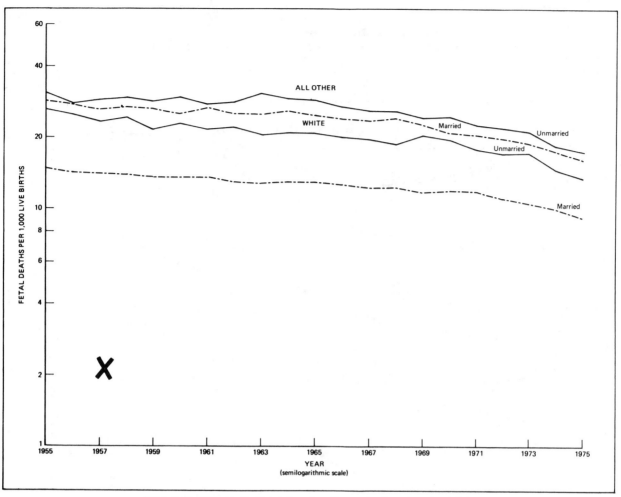

Fetal death ratios by marital status of mother and color: reporting States, 1955-75

Figure 5-16.   A semilogarithmic chart portraying four temporal series of fetal deaths according to the racial and marital status of the mother from 1955 to 1975. This chart is definitely rated unacceptable since it violates several basic design standards. These errors and deficiencies are discussed in the text. (From Stephanie J. Ventura, Trends and Differentials in Births to Unmarried Women: United States, 1970–76, U.S. Department of Health and Human Services, Hyattsville, Md.: National Center for Health Statistics, 1980, p. 33.)

The results of such comparison reveals a marked similarity between the semilogarithmic and arithmetic curves (Figure 5-14 and 5-15). The relatively narrow range covered by the data would seem to account for this similarity. As a general rule, if the range of values of the data is less than 100 percent, or approximately 100 percent, the difference in configuration between the curves plotted on arithmetic and semilogarithmic grids would not be appreciably different. Of course, this assumes comparability in the size and proportions of the two grids.

Figure 5-16, which shows four curves representing trends in fetal deaths according to the marital status and color of the mother, is another example of a poorly designed chart. First, the bottom cycle, with values from 1 to 10, is almost entirely superfluous. Only a very small portion of one curve actually extends into this cycle. Accordingly, the vertical scale (logarithmic) could begin with 8, and the remainder of the cycle could be eliminated. This would provide a wider spread of the four curves, thus reflecting a clearer portrayal of changes and trends. Second, with the elimination of most of the first cycle, the scale figures along with full scale lines and scale points would provide a more precise and more easily interpretable system of values. Third, the curve patterns should be

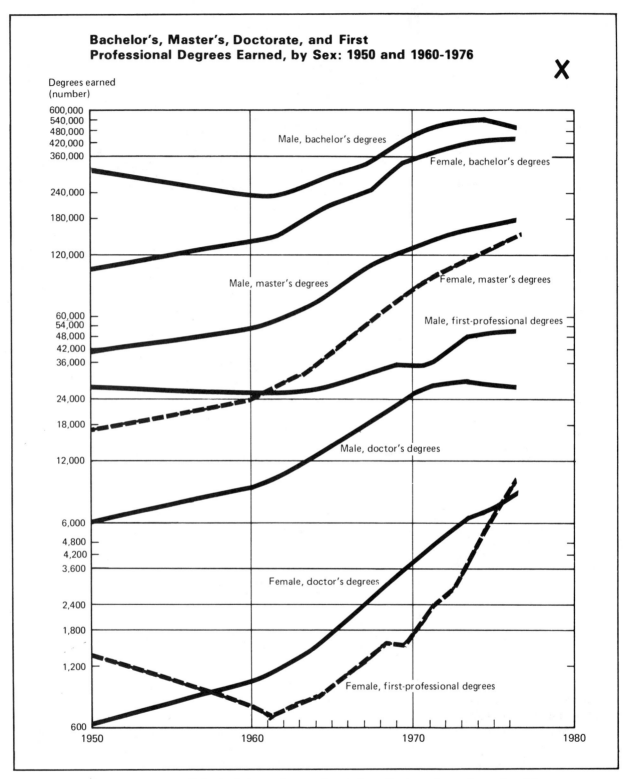

Bachelor's, Master's, Doctorate, and First Professional Degrees Earned, by Sex: 1950 and 1960-1976

X

Degrees earned
(number)

600,000
540,000
480,000
420,000
360,000

240,000

180,000

120,000

60,000
54,000
48,000
42,000
36,000

24,000

18,000

12,000

6,000

4,800
4,200
3,600

2,400

1,800

1,200

600

1950   1960   1970   1980

Male, bachelor's degrees

Female, bachelor's degrees

Male, master's degrees

Female, master's degrees

Male, first-professional degrees

Male, doctor's degrees

Female, doctor's degrees

Female, first-professional degrees

Figure 5-17. A well-drafted chart whose value and quality have been depreciated by idiosyncratic scale figures and scale rulings. See text for more details. (From U.S. Department of Commerce, Office of Statistical Policy and Standards, Social Indicators, 1976, Washington, D.C.: Government Printing Office, 1977, p. 276.)

BACHELOR'S, MASTER'S, DOCTOR'S, AND FIRST PROFESSIONAL DEGREES, CLASSIFIED BY SEX: 1950 AND 1960-1976

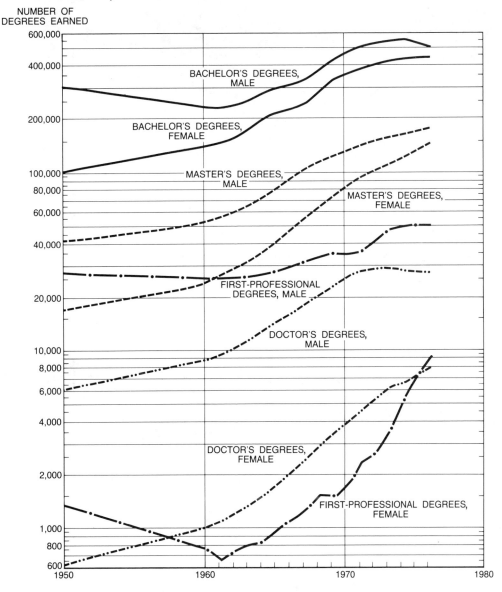

Figure 5-18. **This chart is a revision of Figure 5-17. Note particularly the changes in vertical scale rulings and numerals. The vertical scale now conforms to accepted mathematical standards as well as conventional practice.**

redrawn, and the labels, relocated. Labels are weak and unimpressive. Fourth, the horizontal scale should be recalibrated with full-scale lines and more logical and adequate scale points. Full-scale lines should be drawn at five-year intervals beginning, of course, with 1955. Since the data are plotted on a yearly basis, scale points should be indicated accordingly. In light

of the narrow range and similar trend patterns shown by the four series of data, it probably would have been just as satisfactory to use an arithmetic grid.

At first glance, Figure 5-17 may seem to give the impression of a neat, reliable and well-designed chart, but, unfortunately, the logarithmic scale divisions including scale lines, scale points, and scale des-

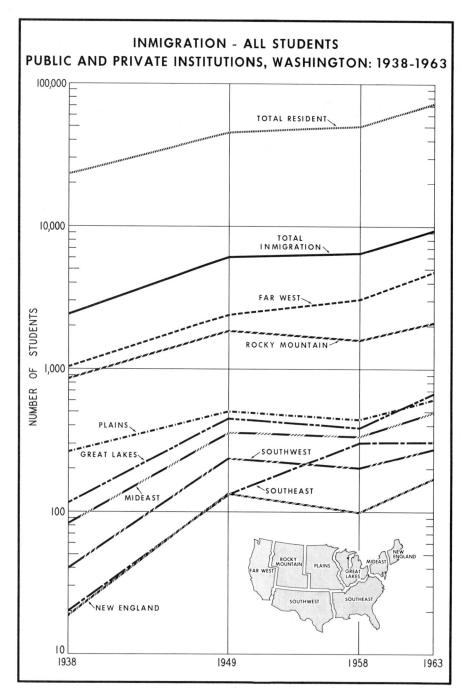

INMIGRATION - ALL STUDENTS
PUBLIC AND PRIVATE INSTITUTIONS, WASHINGTON: 1938-1963

Figure 5-19. This chart indicates how the design of a line chart—semilogarithmic or arithmetic—is adapted to a series of data when there are time intervals of varying lengths. (From Charles S. Gossman, Charles E. Nobbe, Theresa J. Patricelli, Calvin F. Schmid, and Thomas E. Steahr, Migration of College and University Students, State of Washington, Seattle: Washington State Census Board, 1967, p. 39.)

ignations are idiosyncratic and confusing. There is total disregard for the customary divisions that characterize a logarithmic scale along with the consistency and regularity of conventional scale values. As a consequence, the readability and interpretability of the chart have been seriously depreciated. Instead of adhering to the conventional base of ten in determining scale divisions, apparently various multiples of six have been used. Incidentally, in our system of common logarithms, ten is used as the base for determining logarithmic values. Figure 5-17 has been reconstructed in accordance with more acceptable principles of semilogarithmic chart design as shown in Figure 5-18. Note, in particular, that the logarithmic scale has undergone considerable revision. Again, Figure 5-17 illustrates how a single erroneous or eccentric feature in a chart can severely depreciate its quality. Figure 5-18 has been redesigned in accordance with more appropriate and acceptable standards.

SEMILOGARITHMIC CHARTS WITH UNEQUAL TIME INTERVALS

Not infrequently, it may be necessary to construct line charts (semilogarithmic or arithmetic) with data based on irregular time intervals. Figure 5-19 is such an example. The basic data pertain to the interstate migrations of college and university students compiled by the United States Office of Education in 1938, 1949, 1958, and 1963. The particular data presented by Figure 5-19 represent nonresident students attending colleges and universities in the state of Washington. In this chart, residence has been classified by state groupings (regions) as defined by the Office of Education. The small map in the lower right-hand corner of the chart serves as an approximate referent for the eight regions. In addition to separate curves for the respective regions, there is a curve representing the total for all nonresidents and another curve for comparison that shows all the students who are residents of Washington State.

PORTRAYAL OF AGE AND SEX DISTRIBUTIONS OF VITAL AND SOCIAL RATES ON SEMILOGARITHMIC CHARTS

Although the most common use of the semilogarithmic chart is for exhibiting time series, it also may be used for other purposes. For example, sex and age

Figure 5-20. This chart reflects common errors in plotting data according to age intervals. The horizontal scale is calibrated according to five-year intervals but actually only one rate is available for a five-year interval, that for the first five years. Of the remaining nine intervals, eight are 10-year intervals, and the last is open-ended. The plotting of the curves is also incorrect, and the draftsmanship is inferior. (From Kingsley Davis, "Cities and Mortality," Proceedings of the International Union for the Scientific Study of Population, Liège: 1973, pp. 259–281.)

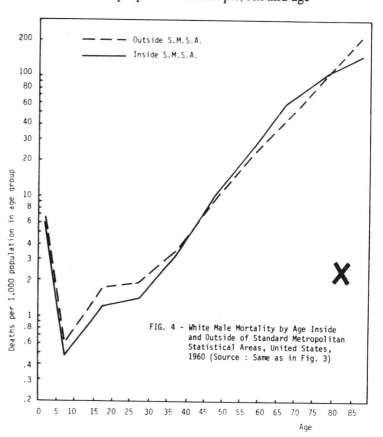

differentials of vital and social rates may be shown effectively by a semilogarithmic chart. Figure 5-20 is an illustration of a series of male mortality rates according to age plotted on a semilogarithmic grid. However, in both design and construction, the chart reflects serious shortcomings. Figure 5-21 is a reconstruction of Figure 5-20. A comparison of the two charts indicates the following changes:

1. The calibrations on the horizontal axis in Figure 5-21 have been redrawn to conform to the age categories indicated by the data, that is, one 5-year interval (0–4); one open-ended interval (85+); and eight 10-year intervals.

2. The actual plotting points are placed at the midpoint of each interval, except in the case of the one open-ended interval (85+), which is fairly arbitrary. It will be observed that the plotting in Figure 5-20 is incorrect.

3. Both vertical and horizontal lines have been added in order to make the chart more accurate and easier to interpret.

4. The two curves have been redrawn so they will stand out more distinctly. Also, their identifying labels have been moved.

5. The title has been redesigned and relocated.

Figure 5-21. A reconstructed version of Figure 5-20. Note especially the plotting of the curves as well as basic design features.

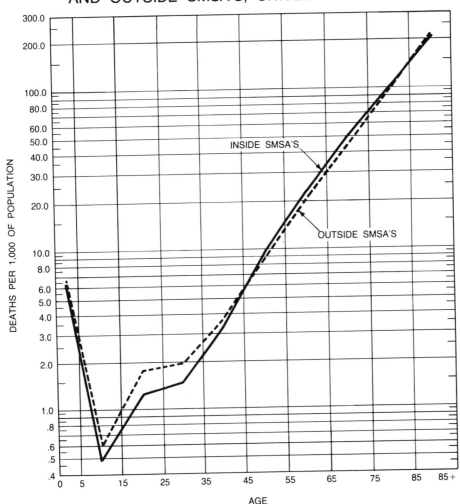

WHITE MALE MORTALITY BY AGE, INSIDE AND OUTSIDE SMSA'S, UNITED STATES: 1960

COMPLETED AND ATTEMPTED SUICIDE, BY SEX

SEATTLE: 1948-1952

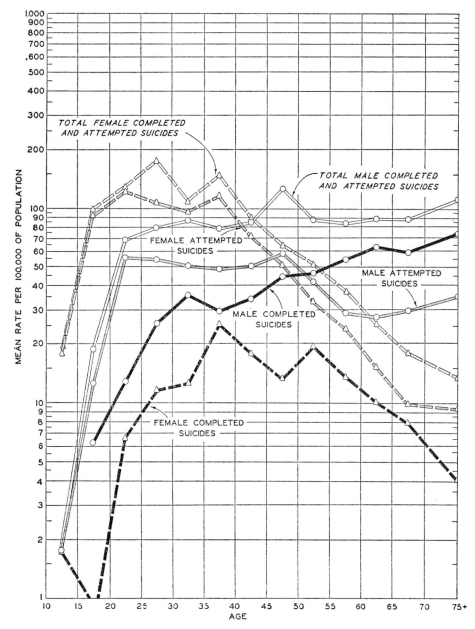

Figure 5-22. Another example of a semilogarithmic chart that has been selected for portraying rates by sex and age. It will be observed that there are six series of rates describing suicidal behavior. Six different curves indicate differentials by sex and age for completed suicide, attempted suicide, and completed and attempted suicide combined. [From Calvin F. Schmid and Maurice D. Van Arsdol, Jr. "Completed and Attempted Suicides: A Comparative Analysis," American Sociological Review, 20 (1955), 273–283.]

CRIME GRADIENTS: OFFENSES
SEATTLE: 1949-1951

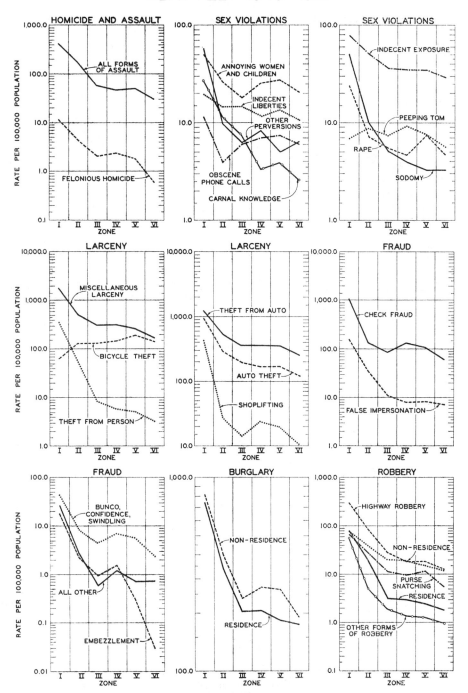

Figure 5-23. *Another example showing the portrayal of nontemporal data on semi-logarithmic grids. The data in all nine panels are spatial data representing crime gradients based on six concentric 1-mile zones radiating from the center (point of highest land value) of Seattle. There is a spatial continuum on the horizontal scale, each division representing one mile. The slope of a curve is indicative of the rate of change from one zone to another. [From Calvin F. Schmid, "Urban Crime Areas: Part II,"* **American Sociological Review,** *25 (1960), 655-678.]*

Figure 5-22 is considerably more complicated than Figure 5-21 in that it depicts six series of data. All the curves in Figure 5-22 pertain to suicidal behavior. In order to obviate confusion as well as to facilitate readability, special attention was given to the six curves. The rationale for differentiating as well as identifying the curves was based on the following specifications: First, the curves representing males are full lines; the curves representing females are dashed lines. Also, for males, the plotting points are circles; for females, triangles. Second, the completed suicide curves are in full black, the attempted suicide curves are stippled, and the combined completed and attempted curves are open. Third, an attempt was also made to use lettering in the differentiation and identification process. Roman type was used to identify

Figure 5-24. Illustration of "constant-dollar" semilogarithmic plotting paper. (Redrawn from Robert E. Sherman, "An Introduction to Constant Dollar Semilogarithmic Plotting Paper," Minneapolis: Hennepin County Office of Planning and Development, 1979.)

HENNEPIN COUNTY BUDGET DATA: 1975-79
(FIGURES IN MILLIONS OF DOLLARS)

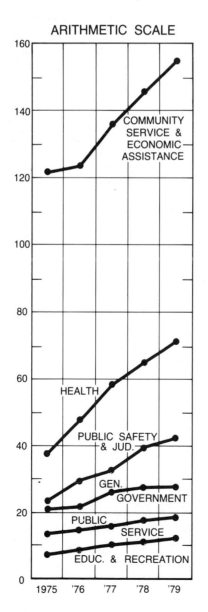

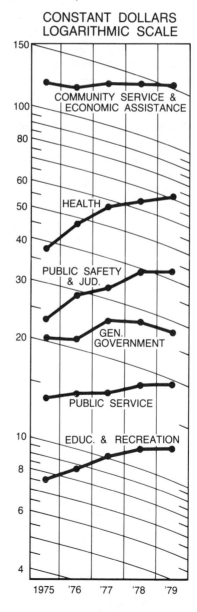

the curves for completed and attempted suicide, and italic type was used for the total completed and attempted suicide curves.

PLOTTING SPATIAL SERIES ON SEMILOGARITHMIC CHARTS

Besides the portrayal of time series and age and sex differentials of vital and social rates, the semilogarithmic chart may sometimes be used for certain spatial series. For example, Figure 5-23 shows a number of crime gradients in a large city. The data represent rates for more than 25 different crime categories, which have been computed for six 1-mile zones radiating from the center of the city. The zones were delineated by a series of concentric circles drawn at 1-mile intervals from the point of highest land value. The semilogarithmic chart makes it possible to portray clearly and accurately generalized spatial configurations of various crimes in the large city. It will be observed that the most prevalent pattern is typified by a relatively high rate in the innermost zone with precipitate declines to the second and/or third zones. The most atypical categories in this respect are bicycle theft and Peeping Tom offenses. In interpreting the chart, the slope of the curves reveals the relative change in rates from one zone to another. In addition, of course, the actual rates can readily be checked by the calibrations and figures indicated on the vertical scale.

CONSTANT-DOLLAR SEMILOGARITHMIC PLOTTING PAPER

Robert E. Sherman, in recognition of the fact that "inflation creates a spurious distortion in the graphic presentation of time sense of dollar-value variables" has devised a special adaptation of semilogarithmic paper that makes possible direct plotting of dollar-value variables adjusted to purchasing power. Conventionally, it is common practice to transform nominal dollar figures into constant dollars by dividing the nominal dollar values for each point in time by the consumer price index. The resultant figures can then be plotted on a rectilinear coordinate or semilogarithmic grid, not infrequently along with the series of nominal dollars. However, Sherman's methodology permits the direct plotting of data in its original, numerical form on the "constant-dollar semilogarithmic plotting paper," thus transforming the series into constant-dollar values. Figure 5-24 illustrates the essential features of this innovation.[7]

---

[7] Robert E. Sherman, "An Introduction to Constant-Dollar Semilogarithmic Plotting Paper," Minneapolis: Hennepin County Office of Planning and Development, 1979. Paper presented at the 1979 Annual Meeting of the American Statistical Association. Constant-dollar plotting paper for the 14-year period 1969–1982 in a two-year semilogarithmic grid is available through Dr. Sherman.

CHAPTER SIX

CHOROPLETH MAPS

Problems, Principles, and Practices

PERHAPS THE BEST KNOWN AND MOST frequently used statistical map is labeled by the cartographer as the "choropleth" (Greek *chōros,* place; and *plethos,* number, quantity) map. Statisticians commonly refer to this type of map as the "crosshatched" or "shaded" map.

Characteristically, the choropleth map portrays range-graded values for specifically delimited areas by means of hatching, shading, or coloring. The areal units that might be represented in a choropleth map may be entire countries, states, counties, census tracts, precincts, or enumeration districts. The statistics shown on choropleth maps are expressed as rates, ratios, percentages, or other statistical measures and indices, not absolute numbers or frequencies. Illustrations of data appropriate for choropleth mapping include rates pertaining to crime, mortality, morbidity, natality, divorce, and marriage; percentages relating to such population characteristics as age, sex, race, nativity, occupation, employment, and educational status, as well as other percentages (e.g., percentage of votes for particular candidates or issues or percentage of land in farms or forests); statistical measures—mean or median rent, mean or median value of dwelling units, or median grade completed for the population 25 years of age and older; indices—per capita income, sales, taxes, per capita consumption of coffee, beer, cigarettes, or sugar; population density; amount of wheat, corn, barley, potatoes, or tobacco produced per acre; and rates of increase or decrease of population, income, sales, manufacturing, or agricultural production.

BASIC ELEMENTS, PROBLEMS AND ERRORS IN CHOROPLETH MAPPING

Although the cartographer and other graphic specialists have long recognized the problems and shortcomings inherent in choropleth mapping, there are still a number of unresolved questions that all too frequently are disregarded or glossed over. In the case of choropleth maps, these questions are greatly complicated by the fact that the graphic specialist has little or no control over certain basic elements in the design process. Although these elements may represent serious constraints, the graphic specialist must attempt to resolve them. Normally, for example, the map designer possesses little or no control over the quality of the data or the size, shape, and relative homogeneity of the areal units that he or she is forced to

use. As a consequence, the shortcomings and idiosyncrasies frequently manifested by choropleth maps may be attributable to the limitations and inadequacies of the elements that the map designer must use rather than to the inappropriateness of the design.

RELIABILITY OF STATISTICAL DATA

Since statistical data are subject to error, it is important to give serious consideration to this fact in the preparation of all kinds of charts. Basic statistical data for choropleth maps are derived either from complete counts or from samples. The actual process of collecting the original data may involve field enumeration or observation, correspondence techniques, or some type of registration. Errors that occur in statistical data may be dichotomized as sampling and nonsampling errors. If, for example, the data are derived from a sample survey, two types of errors are possible—sampling and nonsampling, whereas in the case of a complete census or complete registration, the data are subject only to nonsampling errors.

Nonsampling errors may result from such factors as response variability, response bias, nonresponse, coding and other processing errors, definitional difficulties, undercounting, questionnaire design, and fabrication of data. Sampling errors occur because observations are made only on a sample, not on the entire population. In designing a statistical sample, the size of the sample, the representativeness of the sample, the sampling procedure, and the size of the population or universe from which the sample is derived affect the reliability of the data.

In the United States many of the data utilized in the preparation of choropleth maps, especially in the social sciences, depend either directly or indirectly on decennial census reports of the Bureau of the Census. Generally, since the 1940s, sampling techniques have been used extensively by the Bureau of the Census in their data-gathering operations, including the decennial census. For example, in the 1980 Census, the complete-count items for population include only the following: household relationship, sex, race, age, marital status, and Spanish/Hispanic origin or descent. In addition, there are 24 items collected as sample households. For housing, there are eight complete-count items and 18 sample items, making a total for both population and housing of 14 complete-count items and 42 sample items. Each of the 56 items in the 1980 Census, for example, race, occupation, education, value of home, number of bedrooms, and heating equipment are classified into additional categories.

It should be understood that sample data are only estimates of what a complete count would show. For sample data, the number of cases in specific categories as well as the size of the population or the number of dwelling units in various geographical areas influence the reliability of the data and hence can determine the accuracy and authenticity of a choropleth map. Not infrequently, certain sample data based on enumeration districts, census tracts, and even counties may be too small to provide data sufficiently reliable to portray on a choropleth map.[1]

Table 6-1 illustrates a number of theoretical and practical implications of sampling errors, particularly the role of relatively large and small populations, as well as their special significance in the design and construction of choropleth maps.[2] It will be observed that an understanding of the concepts "standard error" and "confidence limits" is essential for a meaningful interpretation of this table. For those not familiar with the meaning of "standard error" and "confidence limits," the following brief definitions and explanations may be of some help in clarifying these concepts.

The deviation of a sample estimate from the average of all possible samples is called the sampling error. The *standard error* of a survey estimate is a measure of the variation among the estimates from

[1] The problem of statistical error in graphic presentation is discussed in more detail in Chapter 10. In footnotes, figures have been used for chapter references.

[2] Reproduced from Charles P. Kaplan and Thomas L. Van Valey, *Census '80: Continuing the Factfinder Tradition,* United States Bureau of the Census, Washington, D.C.: Government Printing Office, 1980, p. 177.

Table 6-1   Confidence Intervals for Estimates at Different Levels

Occupation (Figures based on 20-percent sample)	Fairfax County, Va.			Tract 4049, Fairfax Co.		
	1970 Census Estimate (1)	95% Confidence interval[1] (2)	Percent Relative Error (3)[2]	1970 Census Estimate (4)	95% Confidence interval[1] (5)	Percent Relative Error[2] (6)
Total employed, 16 years old and over	163,556	162,456 - 164,656	0.7	749	634 - 864	15.4
Professional, technical & kindred workers	48,826	47,926 - 49,726	1.8	256	191 - 321	25.4
Managers & administrators, except farm	22,928	22,258 - 23,598	2.9	107	62 - 152	42.1
Sales workers	13,195	12,695 - 13,695	3.8	78	38 - 118	51.3
Clerical & kindred workers	38,121	37,341 - 38,901	2.1	194	134 - 254	30.9
Craftsmen, foremen & kindred workers	15,841	15,311 - 16,371	3.4	55	20 - 90	63.6
Operatives, except transport	4,454	4,164 - 4,744	6.5	6	([3])	
Transport equipment operatives	3,378	3,128 - 3,628	7.4	0	(NA)	
Laborers, except farm	2,885	2,655 - 3,115	8.0	16	·([3])	
Farm workers	345	265 - 425	23.2	0	(NA)	
Service workers	12,088	11,608 - 12,568	4.0	37	7 - 67	81.1
Private household workers	1,495	1,325 - 1,665	11.4	0	(NA)	

NA Not available
[1] Takes account of sampling variability; range has 95 percent probability of including the value being estimated.
[2] Defined for this purpose as two standard errors as a percentage of the estimate.
[3] Indicates the relative error exceeds 100 percent.
Source: PHC (1)–226 *Census Tracts*, Washington, D.C.–Md–Va SMSA

This table reproduced from Charles P. Kaplan and Thomas L. Van Valey, *Census '80: Continuing the Factfinder Tradition,* United States Bureau of the Census, Washington, D.C.: Government Printing Office, 1980, p. 177.

Geographic Areas in 1970 Census Reports

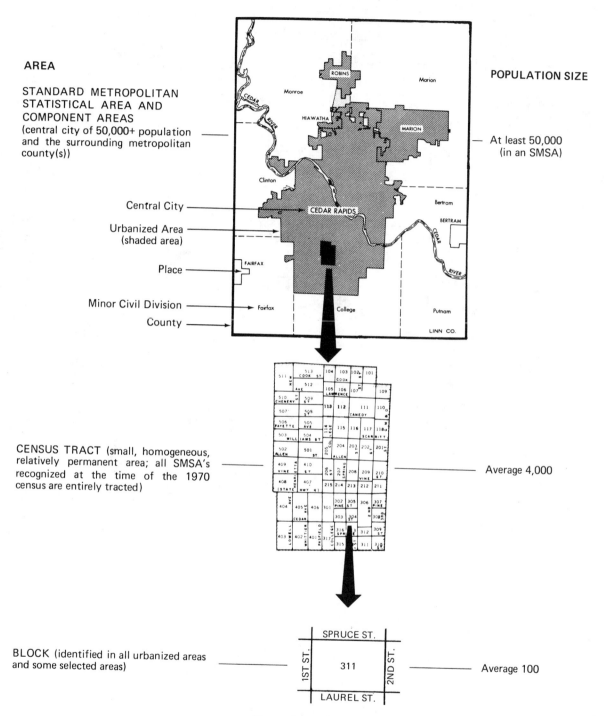

AREA

STANDARD METROPOLITAN STATISTICAL AREA AND COMPONENT AREAS (central city of 50,000+ population and the surrounding metropolitan county(s))

Central City

Urbanized Area (shaded area)

Place

Minor Civil Division

County

CENSUS TRACT (small, homogeneous, relatively permanent area; all SMSA's recognized at the time of the 1970 census are entirely tracted)

BLOCK (identified in all urbanized areas and some selected areas)

POPULATION SIZE

At least 50,000 (in an SMSA)

Average 4,000

Average 100

Figure 6-1. *Illustration of various types of areas used by the United States Bureau of the Census in collecting and tabulating data in connection with censuses and other surveys.* (*From United States Bureau of the Census,* Census Data for Community Action, *Washington, D.C.: Government Printing Office, 1972, p. 4.*)

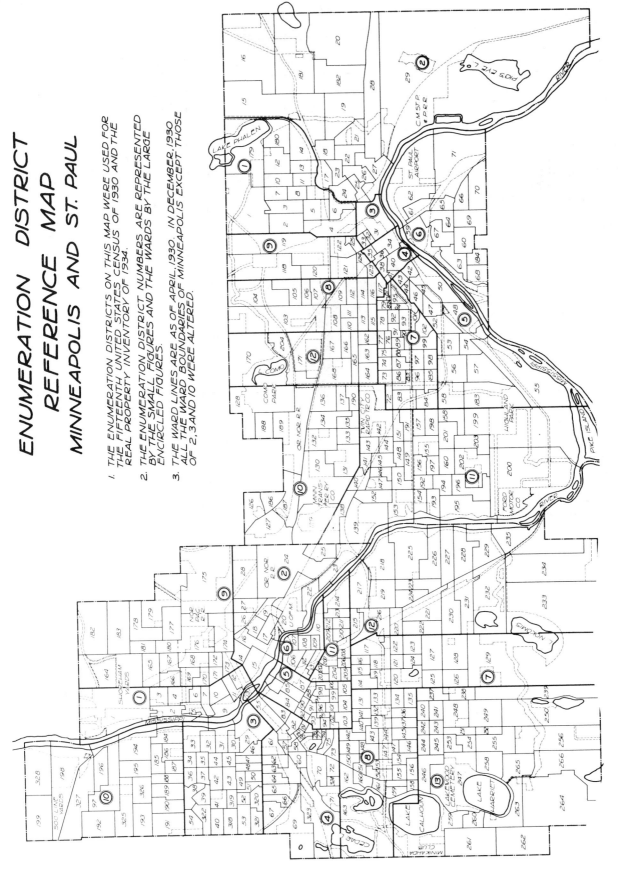

ENUMERATION DISTRICT
REFERENCE MAP
MINNEAPOLIS AND ST. PAUL

1. THE ENUMERATION DISTRICTS ON THIS MAP WERE USED FOR
THE FIFTEENTH UNITED STATES CENSUS OF 1930 AND THE
REAL PROPERTY INVENTORY OF 1934.

2. THE ENUMERATION DISTRICT NUMBERS ARE REPRESENTED
BY THE SMALL FIGURES AND THE WARDS BY THE LARGE
ENCIRCLED FIGURES.

3. THE WARD LINES ARE AS OF APRIL, 1930. IN DECEMBER, 1930
ALL THE WARD BOUNDARIES OF MINNEAPOLIS EXCEPT THOSE
OF 2,3 AND 10 WERE ALTERED.

Figure 6-2. An enumeration district map for two large cities—Minneapolis and St. Paul. Enumeration districts (ED's) are areas used in the geographic control of enumeration activities. An ED is the territory assigned to a single enumerator to cover. With the introduction of mailing techniques in recent censuses, the conventional door-to-door enumeration is used much less than in the past. (From Calvin F. Schmid, Social Saga of Two Cities, Minneapolis: Minneapolis Council of Social

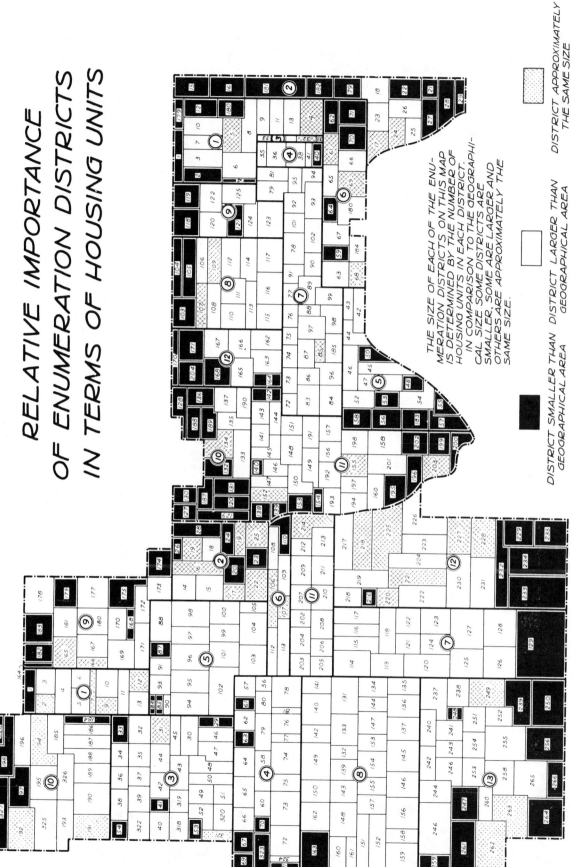

RELATIVE IMPORTANCE OF ENUMERATION DISTRICTS IN TERMS OF HOUSING UNITS

THE SIZE OF EACH OF THE ENUMERATION DISTRICTS ON THIS MAP IS DETERMINED BY THE NUMBER OF HOUSING UNITS IN EACH DISTRICT. IN COMPARISON TO THE GEOGRAPHICAL SIZE SOME DISTRICTS ARE SMALLER, SOME ARE LARGER AND OTHERS ARE APPROXIMATELY THE SAME SIZE.

DISTRICT SMALLER THAN GEOGRAPHICAL AREA

DISTRICT LARGER THAN GEOGRAPHICAL AREA

DISTRICT APPROXIMATELY THE SAME SIZE

Figure 6-3. In choropleth maps, there is a tendency to judge the importance of an area by its territorial size, which, of course, may be misleading. This chart was delineated to indicate the relative importance of each enumeration district in Minneapolis and St. Paul, not in terms of geographical size, but in terms of the number of housing units that each contains. (From Calvin F. Schmid, Social Saga of Two Cities, Minneapolis: Minneapolis Council of Social Agencies, 1937, p. 383.)

the possible samples and thus is a measure of the precision with which an estimate from a particular sample approximates the average result of all possible samples. The *relative standard error* is defined as the standard error divided by the value being estimated. From the sample estimate and an estimate of its standard error, it is possible to construct interval estimates with prescribed confidence that the interval includes the average result of all possible samples. More specifically, with reference to the 95-percent confidence interval indicated in the table, approximately nineteen twentieths of the intervals from two standard errors above the estimate to two standard errors below the estimate would include the average value of all possible samples. The interval from two standard errors below the estimate to two standard errors above the estimate is referred to as a *95-percent confidence interval*.[3]

In addition to the more or less direct utilization of survey and census data in constructing choropleth maps, there are various kinds of derivative data such as rates, indices, and other statistical measures that are applicable for choropleth mapping. Of course, it is essential to give appropriate consideration to their reliability and comparability.[4]

As with the statistics portrayed on choropleth maps, the mapmaker has little or no direct control over the size, shape, and relative homogeneity of the areal units that comprise the map.[5]

As we have noted previously

In the planning and construction of statistical maps it is not only essential to possess a genuine familiarity with the meaning, reliability and sources of data, it is also particularly important to have a clear understanding of the spatial referents used in the compilation of the data. Most of the data used in the construction of statistical maps are based on areas, but the type of area selected may spell the difference between a superior or inferior map.[6]

The following is a brief listing of various kinds of areas used either directly or indirectly in the preparation of choropleth maps: (1) political areas such as states, counties, minor civil divisions, incorporated places, wards, congressional districts, precincts, communes, departments, provinces, and parishes; (2) census tracts that are relatively small and permanently established areas in large cities and their environs that have been laid out for statistical purposes; (3) census county divisions, which are similar to census tracts, but generally cover areas outside large cities and their metropolitan districts; (4) enumeration districts, which are smaller than census tracts and are laid out mainly for administrative purposes for controlling and directing field enumerations; (5) blocks, which are the smallest of these areas and are commonly referred to as city blocks. Figure 6-1 portrays very clearly most of the geographic areas used by the Census Bureau in recent decennial censuses.[7]

---

[3] This paragraph excerpted from Maria E. Gonzalez, Jack L. Ogus, Gary Shapiro, and Benjamin J. Tepping, *Standards for Discussion and Presentation of Errors in Data*, United States Bureau of the Census, Technical Paper No. 32, Washington, D.C.: Government Printing Office, 1974, Appendix 1, p. 1.

[4] In the 1930s there was a notable flurry of discussion concerning the reliability of rates for relatively small geographical divisions. See, for example, Frank Alexander Ross, "Ecology and the Statistical Method," *American Journal of Sociology*, **38** (1933), 507–522; Robert E. Chaddock, "Significance of Infant Mortality Rates for Small Geographic Areas," *Journal of the American Statistical Association*, **29** (1934), 243–249; Frederick F. Stephan, "Sampling Errors and Interpretations of Social Data Ordered in Time and Space," *Journal of the American Statistical Association*, **29**, Supplement (1934), 165–166; Charles C. Peters, "Note on a Misconception of Statistical Significance," *American Journal of Sociology*, **39** (1933), 231–236.

[5] This statement, of course, requires a minor qualification: If files of certain block or individual data are available, it may be possible with the aid of electronic computer equipment to allocate such data in accordance with almost any desired set of areas.

[6] Calvin F. Schmid and Stanton E. Schmid, *Handbook of Graphic Presentation*, New York: John Wiley & Sons, 1979, p. 171.

[7] For a more detailed discussion of areas used in the preparation of statistical maps, see Calvin F. Schmid and Stanton E. Schmid, *Handbook of Graphic Presentation*, New York. John Wiley & Sons, 1979, p. 171–177.

Figure 6-4. In choropleth maps, entire areas representing a particular class-interval value are crosshatched or shaded uniformly regardless of the great differences that may exist within the units. This chart shows that the eight census tracts are far from homogeneous. It also shows that some are relatively more homogeneous than others. The two maps shown in this chart are, of course, choropleth maps based on several hundred city blocks. (From Calvin F. Schmid, Social Trends in Seattle, Seattle: University of Washington Press, 1944, p. 296.)

THE RELATIVE HOMOGENEITY OF
CERTAIN CENSUS TRACTS
A METHODOLOGICAL ANALYSIS

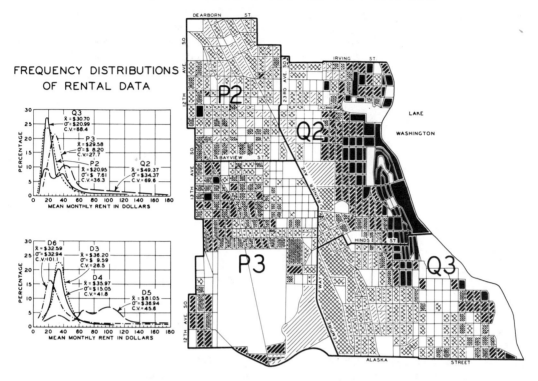

FREQUENCY DISTRIBUTIONS OF RENTAL DATA

N.B. BLOCKS WITH LESS THAN THREE DWELLING UNITS ARE NOT HATCHED.

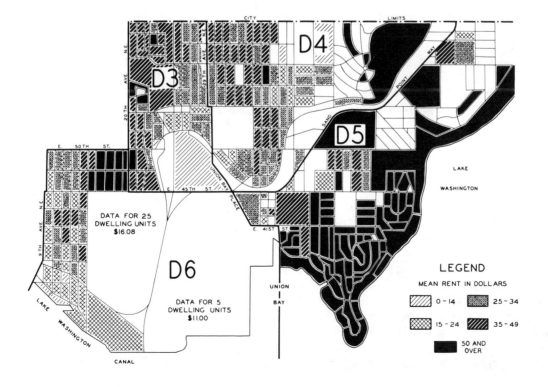

LEGEND

MEAN RENT IN DOLLARS

▨	0 – 14	▨	25 – 34
▨	15 – 24	▨	35 – 49
■	50 AND OVER		

Figure 6-2 shows how the cities of Minneapolis and St. Paul were subdivided into 452 enumeration districts for the decennial population census of 1930 and the special real property inventory of 1934.[8] Fortunately, the relatively small, homogeneous enumeration districts, along with detailed 100-percent population and housing data, presented an unusually favorable combination for the preparation of a series of choropleth maps. At the present time, and for the past few decennial censuses, reliable maps of this kind would be impossible to prepare since most of the population and housing items are compiled on a sample basis. Sample data such as those obtained in decennial censuses for enumeration districts would not be sufficiently reliable to warrant the preparation of choropleth maps.[9]

INHERENT SHORTCOMINGS OF CHOROPLETH MAPS ATTRIBUTABLE TO AREAL UNITS

Although as a medium of visual communication the choropleth map embodies many valuable features, there are three specific shortcomings frequently attributable to the inherent rigidity, artificiality, and illogicality of areal units, perhaps the most characteristic and essential component of the choropleth map.

First, the amount of cross-hatching or shading is determined solely by geographic area and not by the number of cases or size of population each areal unit contains. Visually, certain units may seem very important merely because of the relatively large area they comprise, but actually, in terms of the number of cases or size of population each areal unit contains, they may rank very low. Figure 6-3 was delineated to indicate the relative importance of each enumeration district in Minneapolis and St. Paul, not in terms of geographical size, but in terms of the number of housing units that each contains. A comparison of Figures 6-2 and 6-3 clearly reveals that some districts which normally cover a large geographical area are relatively small and unimportant in terms of the number of housing units, while other districts that are very small geographically are really much more important in terms of the number of dwelling units which they contain.[10]

Second, entire areal units on choropleth maps representing particular class-interval values are crosshatched or shaded uniformly regardless of the great differences that may exist within the units. For example, one small corner of an area may be densely populated, whereas other parts may be relatively sparse or entirely uninhabited, but the entire area is portrayed as an average with uniform shading. Also, an areal unit may be heterogeneous with respect to any number of indices, but on a choropleth map it would be most difficult, if not impossible, to indicate clearly any differentation from one part of the area to another. Accordingly, the entire area is shown uniformly with one type of hatching or shading. Generally, the larger the area, the greater the heterogeneity. Entire states or countries are extreme examples of this anomaly. However, relatively small areas are frequently far from being homogeneous.

For a clear illustration of this fact the reader is referred to Figure 6-4, which summarizes in part the results of an empirical analysis of eight census tracts in two different sections of a large city. The eight tracts have been broken down into over 700 blocks with mean monthly rent indicated for each block. It is obvious from an examination of Figure 6-4 that these eight tracts are not uniformly homogeneous. For example, tract D3 shows a mean monthly rent of $36.20, a standard deviation of $9.59, and a coefficient of variation of 26.5 percent. By contrast, tract D6 has a mean monthly rent of $32.59, a standard deviation of $32.94, and a coefficient of variation of 101.0 percent. Other tracts in Figure 6-4 indicate similar variable patterns.[11]

Third, since data for choropleth maps are based on discrete areal divisions, an impression of definitive and abrupt change may be conveyed by maps of this kind. In reality, however, the transition in value from one spatial unit to another is usually gradual and not abrupt. Incidentally, in order to adjust for the abrupt changes in value that characterize choropleth maps, the mapmaker may choose to shift the basic design to an isopleth map.

---

[8] Census tracts in Minneapolis and St. Paul were first used by the United States Bureau of the Census in 1940. They were laid out by the present writer and officially approved in December 1934. At that time, there were 121 census tracts in Minneapolis and 76 in St. Paul, less than one half of the 452 enumeration districts used in 1930 and 1934.

[9] Calvin F. Schmid, *Social Saga of Two Cities*, Minneapolis: Minneapolis Council of Social Agencies, 1937, passim.

[10] Calvin F. Schmid, *Social Saga of Two Cities*, Minneapolis: Minneapolis Council of Social Agencies, 1937, pp. 381–386.

[11] For a more complete discussion of the ecological, statistical, and graphical implications of the homogeneity of areal units, see Calvin F. Schmid, *Social Trends in Seattle*, Seattle: University of Washington Press, 1944, pp. 294–297.

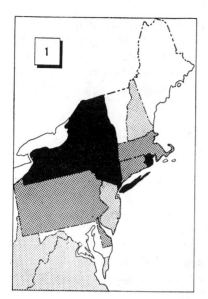 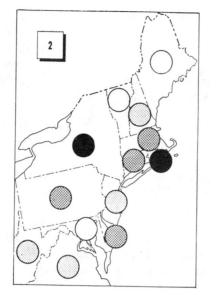 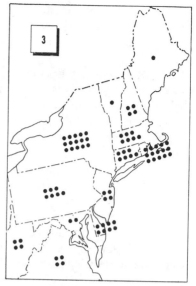

Figure 6-5. *Two suggested solutions for "area bias" of choropleth maps. Panel 1 is a conventional choropleth map. In Panel 2, the several states are represented by a series of circles of identical size and shaded to match the intervals used in Panel 1. In Panel 3, the value intervals are represented by a varying number of black dots. [From Kenneth W. Haemer, "Area Bias in Map Presentation," The American Statistician, 3 (April–May 1949), 19.]*

SUGGESTED SOLUTIONS FOR AREA BIAS IN CHOROPLETH MAPPING

As indicated in the preceding section, there is a tendency to judge the importance of areas on a choropleth map on the basis of territorial size rather than the number of cases or the size of the population that they represent. Several solutions have been proposed to remove or at least minimize area bias of this kind.[12]

It will be observed from Figure 6-5 that Panel 1 represents a conventional choropleth map where each state in the northeast portion of the United States is completely shaded in accordance with certain predetermined values. Panel 2 portrays a series of two-dimensional symbols of uniform size, each symbol representing a state. The shading of the symbols corresponds to the shading in Panel 1. Again, in order to circumvent "area bias," a series of dots are superimposed on the map in Panel 3 to represent certain values for the several states. As a consequence of the circle and dot systems in Panels 2 and 3, the "small state" problem has resulted, since it is not possible to superimpose in the smaller areas symbols of this kind or number without overlap. However, where this overcrowding occurs, the symbols have been placed adjacent to the smaller states.[13]

A third proposal to remove "surface-area" bias of the choropleth map is the "three-factor" circle. This technique is illustrated by Figure 6-6. Besides removing "surface-area" bias, it also permits the simultaneous consideration of two pertinent factors instead of one. The "three factor" circles show the number of dwellings (indicated by size of circles), type of residence (shown by position of sectors of circles—urban, rural nonfarm, and rural farm) and percentage of dwellings that have a radio (shading of sectors).[14]

Another solution for overcoming area bias is to redesign the base map so that each area unit is proportional to some value more pertinent or representative than mere territorial size. This type of map has been labeled variously as the "proportional," "distorted," or "value-by-area" map. Two examples of this type of map are shown in Figure 6-7. The base map for this chart was drawn in proportion to the

---

[12] It should be noted that, if the index or other mensurational unit on a choropleth map involves area such as the number of inhabitants per square mile, its portrayal without correction or adjustment is appropriate.

[13] Kenneth W. Haemer, "Area Bias in Map Presentation," *The American Statistician*, **3** (1949), 20.

[14] Homer L. Hitt, "Three-Factor Cartographic Representation," *The American Statistician*, **2** (1948), 21–22.

1940 population of each state. The upper map depicts the percentage of blacks in each state, and the lower map, the proportion of the population classified as "urban."

A fifth method for correcting or avoiding area bias is to utilize symbols of varying size. These symbols may be one-dimensional (bars and columns), two-dimensional (circles and squares), or three-dimensional (spheres and cubes). The sizes of the symbols superimposed on the map indicate the values they represent. Figure 6-6 is a special variation of the two-dimensional symbol for maps of this kind. Although Figure 6-6 possesses certain distinctive features, the symbols are two-dimensional and therefore this figure may be included as an example of the fifth type of solution. Figure 6-8 is another illustration in which one-dimensional symbols in the form of col-

umns in oblique projection are superimposed on a map. The height of each column is commensurate with the robbery rate in each of the 16 areas (Planning and Administration Area 4 has been subdivided into its four constituent counties).

GROUPING THE DATA INTO CLASS INTERVALS

Frequently, one of the most important and difficult steps in constructing a choropleth map is organizing the data into class intervals. There is no consensus among cartographers and other specialists concerning the most reliable and effective procedure for determining class intervals. Perhaps no single "best" method exists since there is so much variation in the

Figure 6-6.   Another proposal for solving "area bias" in choropleth maps. The number of dwellings in each state is indicated by the size of circles. The percentage of dwellings with radios, in accordance with a threefold classification of type of residence, is shown by a cross-hatching scheme. [From Homer L. Hitt, "Three-Factor Cartographic Representation," The American Statistician, 2 (February 1948), 21–22.]

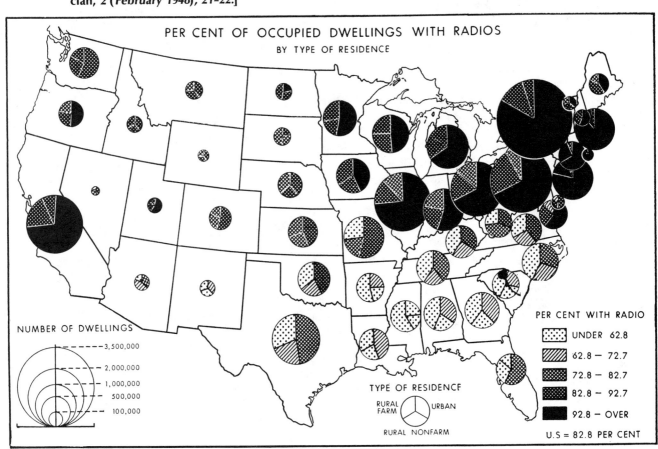

EACH STATE IS REPRESENTED
ON THE SCALE OF ITS POPULATION

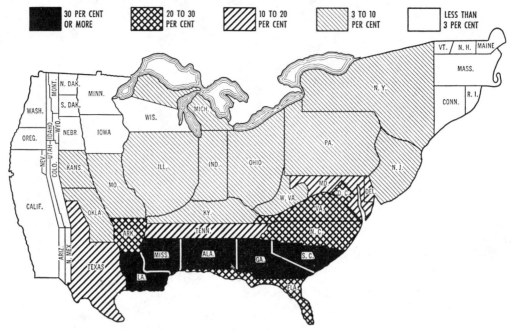

NEGROES IN THE UNITED STATES: PERCENTAGE IN EACH STATE, 1940

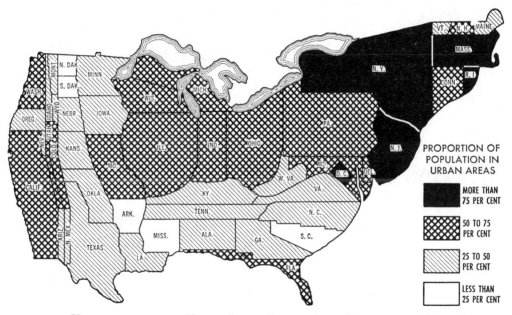

URBANIZATION IN THE UNITED STATES: PROPORTION OF POPULATION IN URBAN AREAS IN
EACH STATE, 1940

Figure 6-7. A fourth suggested solution for "area bias" in choropleth maps is the "proportional," "distorted," or "value-by-area map." In this type of map the size of the areal units is proportional to the data represented, rather than geographical size. For example, in this chart, both maps are scaled in terms of the size of population of each state. (From W. S. Woytinsky and E. S. Woytinsky, World Population and Production: Trends and Outlook. Copyright 1953 by the Twentieth Century Fund, Inc. Reproduced by permission.)

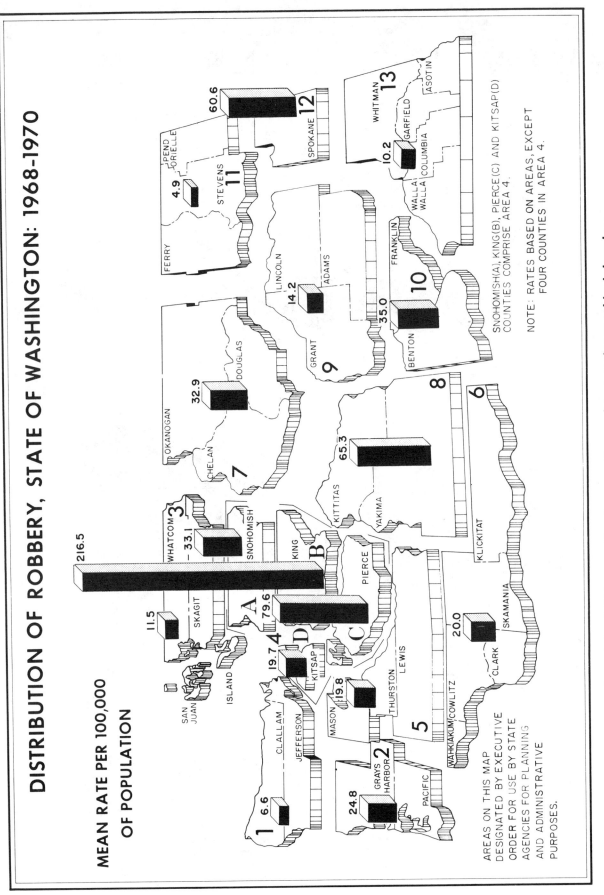

DISTRIBUTION OF ROBBERY, STATE OF WASHINGTON: 1968-1970

MEAN RATE PER 100,000
OF POPULATION

SNOHOMISH(A), KING(B), PIERCE(C) AND KITSAP(D)
COUNTIES COMPRISE AREA 4.

NOTE: RATES BASED ON AREAS, EXCEPT
FOR COUNTIES IN AREA 4.

AREAS ON THIS MAP
DESIGNATED BY EXECUTIVE
ORDER FOR USE BY STATE
AGENCIES FOR PLANNING
AND ADMINISTRATIVE
PURPOSES.

Figure 6-8. A fifth suggestion for correcting or circumventing area bias is based on the substitution of symbols of different sizes. The symbols may be one-, two-, or three-dimensional. Two-dimensional symbols in their most simplified form are used in Figure 6-6. The present chart uses one-dimensional symbols in the form of columns drawn in oblique projection. (From Calvin F. Schmid and Stanton E. Schmid, Crime in the State of Washington, Olympia: Law and Justice Planning Office, Planning and Community Affairs, 1972, p. 49.)

type and quality of data, number, size, and other characteristics of areal units; purpose or objectives in preparing a map; and background, orientation, and understanding of prospective users of a map.

As far as actual practice is concerned, George F. Jenks and Michael R.C. Coulson remark that

> An analysis of maps prepared by authors in various academic disciplines fails to show any rational or standardized procedure for the selection of class intervals. Evidently intuition, inspiration, revelation, mystical hunches, prejudice, legerdemain, and predetermined ideas of what the class intervals should be have characterized the work of most map-makers.[15]

It has been suggested that in order to develop greater insight into statistical mapping, particularly with respect to choropleth and isopleth maps, a genuine grasp of three interrelated phases of the map planning process is essential: (1) the concept of the statistical surface, (2) generalizations resulting from the selection of numbers of classes, and (3) generalizations that depend on the mathematics of classing data.[16]

First, the statistical surface of a choropleth map may be portrayed as a series of tangential prisms, the height of each prism varying in accordance with the values represented by the various unit areas.[17] The choropleth map is a two-dimensional representation of three-dimensional phenomena. Second, the number of intervals chosen for classing the data on a choropleth map determine the degree of generalization of the data as reflected by the map. For example, it is obvious that the degree of generalization of data that are divided into two intervals is considerably less than the degree of generalization where the data are grouped into seven or eight intervals. Third, the method used in classing the data exerts a substantial influence on the form or pattern of generalization indicated on a choropleth map. Although the number of intervals may be identical, the specific values for each interval may be very different depending on the particular technique used in classing the data. In summary, the judgments made by the map-maker concerning his or her personal concept of the statistical surface, of the most desirable degree of generalization, and of the process selected for classing the data control and shape a generalized statistical surface, which is then symbolized to represent the abstract data.[18]

How should the class intervals be determined?

Many methods have been used to select intervals for choropleth maps. They range from those that may be described as superficially impressionistic and arbitrary to those that represent fairly sophisticated mathematical procedures. In a recent paper on the selection of class intervals, Ian S. Evans briefly describes 16 different rationally based techniques.[19]

A fundamental prerequisite in selecting as well as applying any technique is to acquire a clear understanding of the basic data including their distributional properties. In this connection, the preparation of an array or frequency distribution with small equal class intervals is essential. An array or frequency table of this kind will not only provide an opportunity to study the distributional characteristics of the data, but also establish a sound basis for further analysis.

Experience indicates that spatial series of data do not generally conform to symmetrical bell-shaped distributions. Many series are markedly skewed, multimodal, or U-shaped or J-shaped in varying degrees. Figure 6-9 portrays 36 curves representing the distribution of 18 housing variables based on over 260 enumeration districts in Minneapolis and over 180 enumeration districts in St. Paul.[20]

---

[15] George F. Jenks and Michael R. C. Coulson, "Class Intervals for Statistical Maps," *International Yearbook of Cartography,* **III** (1963), 119–134.

[16] George F. Jenks, "Generalization in Statistical Mapping," *Annals, Association of American Geographers,* **53** (1963), 15–26.

[17] The first empirical and theoretical discussion of the statistical surface concept as applied to choropleth and isopleth maps will be found in Calvin F. Schmid and Earle H. Mac Cannell, "Basic Problems, Techniques, and Theory of Isopleth Mapping," *Journal of the American Statistical Association,* **50** (1955), 220–239. Also, in this connection it will be noted that, "as a graphic device, the smooth or stepped surface is surely one of the most dramatic and interesting cartographic methods for portraying quantitative data, as is borne out by the growing use of surfaces in place of other maps in the popular press, in scientific journals, and in atlases." David J. Cuff and Kenneth R. Bieri, "Ratios and Absolute Amounts Conveyed by a Stepped Statistical Surface," *The American Cartographer,* **6** (1979), 157–168.

[18] George F. Jenks, "Generalization in Statistical Mapping," *Annals, Association of American Geographers,* **53** (1963), 15–26.

[19] Ian S. Evans, "The Selection of Class Intervals," *Transactions: Institute of British Geographers,* New Series, **2** (1977), 98–124.

[20] Calvin F. Schmid, *Social Saga of Two Cities,* Minneapolis: Minneapolis Council of Social Agencies, 1937, pp. 381–386.

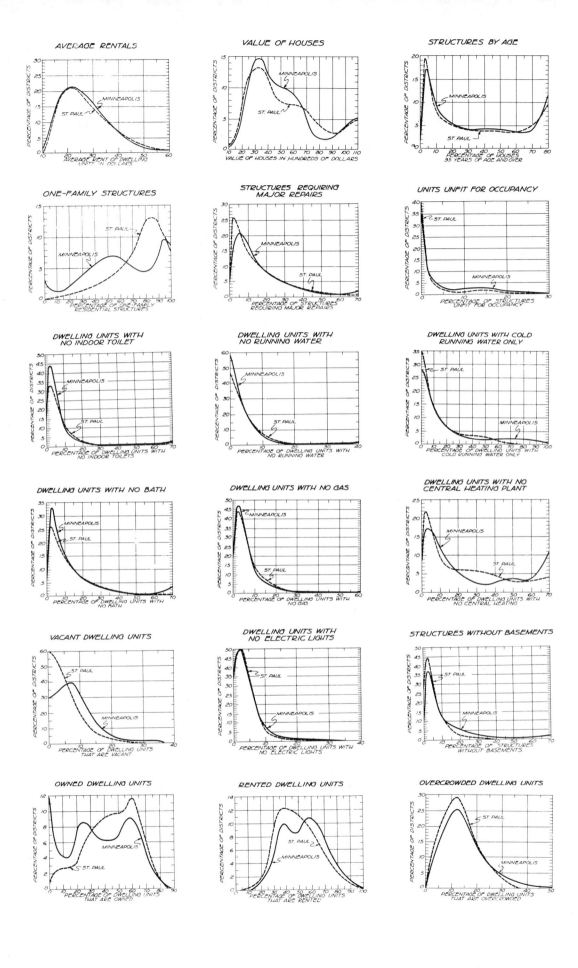

When a mapmaker is faced with the task of designing choropleth maps involving 18 statistical series, such as those portrayed in Figure 6-9 for two large cities with a total of approximately 450 areal units, the procedure selected for determining class intervals is of utmost importance. Moreover, rounding out the total context in which the method for determining class intervals was chosen, the following specifications were also included: (1) The 18 maps should embody sound, meaningful, and significant generalizations; (2) the 18 maps should be reasonably consistent and comparable; (3) the maps should be relatively simple and easy to interpret; and (4) the maps should appeal to the user, not discourage or frustrate him or her.

In order to achieve or at least approximate these goals effectively, it was decided that a relatively simple but thoroughly rational procedure would be most appropriate. A careful study of "natural breaks" and other characteristics of each of the 36 arrays, along with even-numbered designations for the class intervals were followed. More sophisticated procedures such as quantile measures, standard scores, "nested means," standard deviation, iterative methods, or arithmetic progressions, would have posed certain limitations and disadvantages depending on the particular method selected.[21] For example, these limitations and disadvantages could include: (1) several, if not all, intervals being unequal; (2) designation of intervals by odd numbers, including decimals; (3) lack or diminution of comparability among the 18 maps; and (4) diminution of the series' appeal to the average user (because of increased complexity and difficulty in interpretation). In order to obviate any erroneous inferences, the presentation in this paragraph and the preceding one pertains to the 18 maps under discussion, not to all choropleth maps. Certainly, there is no intention of arguing against the use of rigorous and sophisticated mathematical procedures, providing, of course, they are appropriate to the problem at hand. The particular method that is selected should be determined by the various factors and circumstances presented by each special problem as they may occur.[22]

CHOROPLETH MAPS WITHOUT CLASS INTERVALS?

Since it is "now technologically feasible to produce virtually continuous shades of gray by using automatic map drawing equipment," W. R. Tobler suggests that "it is therefore no longer necessary for the cartographer to 'quantize' data by combining values into class intervals."[23] He asserts further that, since the visual intensity of the shading or hatching can be made exactly proportional to the data intensity, the difficult problem of optimum class intervals is circumvented, and the quantization error is nonexistent. Tobler does point out that the main argument in favor of using class intervals seems to be that their use enhances readability. So far, the general response of cartographers to Tobler's no-class mapping technique has been less than enthusiastic.

In a critique of this proposal, Michael W. Dobson avers that Tobler "mistakenly relates quantization of data only with technological feasibility and relegates the ability of the viewer to discriminate between tones to the status of a problem of map generalization. . . . A plotter can draw 105 patterns with visual intensities exactly proportional to the data intensity. "These would, of course, be superimposed on the borders of the 105 enumeration cells. Obviously, this procedure would render the problem of class inter-

[21] A good discussion of most of these methods may be found in Arthur Robinson, Randall Sale, and Joel Morrison, *Elements of Cartography,* New York: John Wiley & Sons, 1978, pp. 171–180 and 410–413.

[22] A computer program for eight classification methods will be found in Kang-Tsung Chang, "An Instructional Computer Program on Statistical Class Intervals," *The Canadian Cartographer,* 11 (1974), 69–77.

[23] W. R. Tobler, "Choropleth Maps Without Class Intervals?" *Geographical Analysis,* 5 (1973), 261–265.

Figure 6-9.   The 18 sets of frequency curves represent 18 housing variables based on approximately 450 enumeration districts in Minneapolis and St. Paul. It is apparent that the spatial distribution of housing variables manifest great irregularity. All the curves are skewed, and the configuration of some are J-shaped or U-shaped. (From Calvin F. Schmid, Social Saga of Two Cities, Minneapolis: Minneapolis Council of Social Agencies, 1937, pp. 208–210.)

vals vacuous. Unfortunately, it would also render the choroplethic form of map vacuous."[24]

A more substantial assessment of the feasibility, advantages, disadvantages, and possible applications of the ideas presented in this section probably should await additional experience and empirical analysis.[25]

HATCHING AND SHADING PROBLEMS

The principle of contrast, so widely applied in statistical graphics, is used freely and extensively in choropleth maps either through the application of color (hue, value, and chroma) or pattern and texture (cross-hatching) and/or value (gray-scale density).[26] The quality of hatching or shading that is used on monochromatic choropleth maps is of primary importance in judging their accuracy, clarity, interpretability, unity, and other essential characteristics as an effective medium of visual communication.

Cross-hatching or shading for choropleth maps may be drawn manually with pen (sometimes brush or airbrush) and ink by a draftsman (drafter), printed mechanically by electronic computer and auxiliary equipment, or cut out and mounted from commercially preprinted self-adhesive sheets. All three types of cross-hatching or shading possess certain advantages as well as disadvantages.

Because of their extensive use, general familiarity, ready availability, and large selection of patterns, the following discussion will be presented mainly within the context of commercially printed hatching and shading material. An examination of preprinted hatching and shading sheets will reveal three basic styles and textures: (1) evenly spaced dots of many sizes and densities, (2) evenly spaced lines of varying weights and widths, and (3) irregularly shaped and frequently irregularly spaced dots and lines. Evenly spaced dots and evenly spaced patterns are most commonly used on choropleth maps.

Obviously, it is of crucial importance in the preparation of a monochromatic choropleth map to select a series of shadings or hatchings for the several class intervals that provide a sequence of visual impressions commensurate with the mensurational values represented. The theoretical principle for judging the gradation of shadings in a choropleth map is based on the gray scale or gray spectrum concept. The gray scale or gray spectrum ranges in value from pure white to total black. It can be described by the black/white ratio, which indicates the percentage of an area that is inked. Psychophysicists and psychologists have demonstrated with considerable certainty that human perception of density of shading as expressed by black/white ratios does not conform to a linear function. That is to say, human beings do not perceive increased density of tone in a direct relationship with the proportion of an area that is inked. Beginning with the Weber–Fechner law promulgated many years ago and subsequently by the work of A.E.O. Munsell, Wilhelm Ostwald, S.S. Stevens, Aeronautical Chart and Information Center, and Robert L. Williams, the relationship of human perception and density of shading represents some kind of curvilinear function.[27]

Although several hundred preprinted self-adhesive shading and hatching patterns are available commercially, the actual selection process for a choropleth map can turn out to be extremely complicated and time-consuming. Similarly, the problems involved in the selection process would not necessarily be any less difficult if the shading or hatching were prepared manually by a draftsman or generated by an electronic computer. The patterns that are selected should present a clearly perceived gradation of tone

---

[24] Michael W. Dobson, "Choropleth Maps Without Class Intervals?: A Comment," *Geographical Analysis,* 5 (1973), 358–360. Also, see: George F. Jenks, "Optimal Data Classification for Choropleth Maps," Occasional Paper 2, Lawrence: Department of Geography, University of Kansas, 1977.

[25] The findings of the following studies pertaining to the response of readers to continuously shaded choropleth maps will be found suggestive, but not conclusive. Jean-Claude Muller, "Perception of Continuously Shaded Maps," *Annals of the Association of American Geographers,* **69** (1979), 240–249; Michael P. Peterson, "An Evaluation of Unclassed Cross-Line Choropleth Mapping," *The American Cartographer,* **6** (1979), 21–37.

[26] Because of their special problems and ramifications, colored choropleth maps are beyond the scope of the present chapter. Coverage will be restricted to monochromatic choropleth maps.

[27] For a more detailed discussion of this question, including the citation of a number of pertinent references, see George F. Jenks and Duane S. Knos, "The Use of Shading Patterns in Graded Series," *Annals of the Association of American Geographers,* **51** (1961), 316–334. In addition to the paper by Jenks and Knos, as well as their citations, a more recent paper by Jon A. Kimerling will be found useful: "A Cartographic Study of Equal Gray Scales for Use with Screened Gray Areas," *The American Cartographer,* **2** (1975), 119–27.

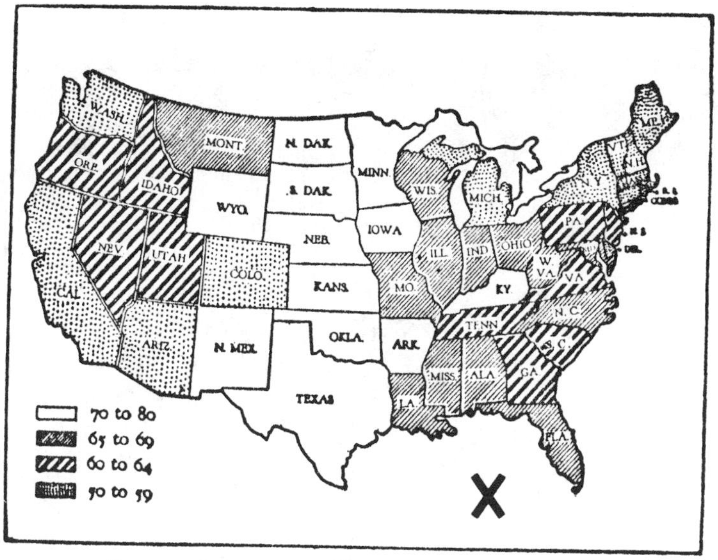

Per cent of drafted men passing physical examination, by States.

Figure 6-10. A poorly designed choropleth map. Although the map is relatively simple with only four class intervals, the hatching scheme follows no gradation. The map is most ineffective and confusing. Failure to adjust for reduction has further depreciated the map. (From Leonard P. Ayres, The War With Germany: A Statistical Summary, Washington, D.C.: Government Printing Office, 1919, p. 17.)

density from light to dark or vice versa with an appearance of fairly distinct steps corresponding to the several class intervals in the map. Dot patterns, especially fine-textured dot patterns seem to be preferred by map users. However, line patterns or a combination of line and dot patterns are used just as frequently on choropleth maps. In actual practice, it may be difficult to find six or seven readily discriminated dot patterns of a particular scale. Also, in this connection it should be clearly recognized that the textural qualities of a pattern influence the perceived density of tone, apart from its black/white ratio.

In general, the number of class intervals on a choropleth map is obviously related to the time and effort involved in constructing a suitable shading or hatching scheme, one that is well-defined and readily discriminable. As previously implied, there is no consensus concerning the optimal number of class intervals on choropleth maps. Usually, there are no fewer than four, nor more than seven or eight, class intervals. Sometimes under special circumstances there may be less than four intervals and possibly as many as twelve. Even in the case of excellently constructed shading or hatching schemes with more than seven or eight categories, perception studies have shown that it is very difficult, if not impossible, to discriminate among the patterns with an acceptable degree of accuracy. However, it is not possible to state precisely what the threshold of discrimination is for shaded and hatching patterns. Experience suggests that seven or eight are maximal.

The size or scale of the original map along with the amount of reduction when it is reproduced are of crucial importance in planning a shading scheme for a choropleth map. Many of the commercially preprinted, self-adhesive patterns are ill-adapted to certain maps because of incompatibility in size. That is, the base map may be either too small or too large for certain preprinted patterns. Furthermore, even when a seemingly appropriate shading or hatching scheme has been carefully prepared and superimposed on a choropleth map, the normal printing process involving reduction and reproduction may create distortions and/or deficiencies in the shading patterns. For example, as a consequence of reduction, some of the dots or lines could fade out or the inked portion could coalesce to such an extent that it would distort or destroy the essential features of certain patterns. Before a final selection is made for a particular series of shadings or hatchings, it may be found expedient to subject them to photographic or photostatic reduction.

EXAMPLES OF DEFICIENT CHOROPLETH MAPS: A CRITICAL EVALUATION

Figures 6-10 and 6-11 are good illustrations of how not to design choropleth maps. Their most glaring deficiences are manifested by the hatching schemes. Both maps do violence to a basic principle of choro-

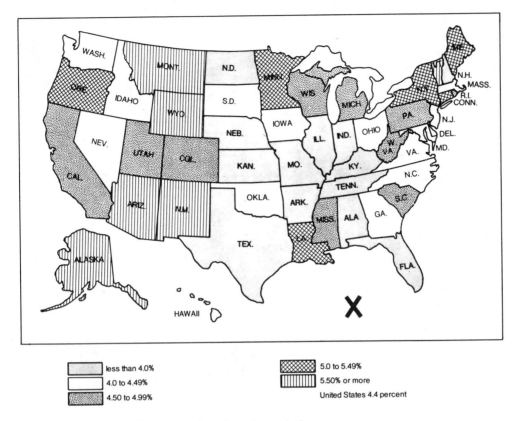

Figure 6-11. *Another example of a choropleth map with an inconsistent and confusing shading scheme. There should have been a visual gradation in the shading from light to dark. (From National Center for Education Statistics, HEW, Selected Statistical Notes on American Education, Washington, D.C.: Government Printing Office, 1975, p. 5.)*

pleth mapping, which is that there should be an increased or decreased density of tone graded in accordance with the values the hatching or shading scheme is supposed to represent. Figure 6-10 shows for the 48 states the percentage of draftees who survived two physical examinations in World War I. The first examination was given by local boards and the second by army physicians after the group of nonrejectees by local boards were sent to camp. In designing the map, the percentage of draftees who were finally accepted into military service were grouped into four intervals ranging from 50 to 80 percent. Figure 6-11 shows a comparison of expenditures for public elementary and secondary day schools for the 50 states based on an index expressed as a percentage of school expenditures to personal income. It will be observed that some states spent less than 4.0 percent of their personal income on schools, while other states spent more than 5.5 percent. Presumably, according to the legend in Figure 6-10, the shading gradation

seems to be from dark to light, but it is difficult to perceive a precise order, except, of course, that the highest category is white (unshaded or unhatched). In Figure 6-11, it is more obvious that no order at all

---

[28] Ironically, the source of this chart, Figure 6-10, has been cited a number of times for the exemplary quality of its graphic material. For example, one writer states that, "It is probably one of the best graphic works done in this country up to that time." Paul J. Fitzpatrick, "The Development of Graphic Presentation of Statistical Data in the United States," *Social Science,* **37** (1962), 203–214. Another writer indicates that this volume "contains some of the best graphic work done in the United States." H. G. Funkhouser, "Historical Development of the Graphical Representation of Statistical Data," *Osiris,* **3** (1973), 269–404. Judged on the basis of present-day standards, the charts in Leonard P. Ayres, *The War with Germany: A Statistical Summary,* are generally acceptable. None are "distinguished" or "outstanding," and the map under discussion is perhaps the "poorest" chart in the entire report.

exists in the shading scheme. In both maps, such confusion and ambiguity in the shading system is totally unprofessional. Furthermore, it should be pointed out that in drafting the original map shown in Figure 6-10, the delineator failed to adjust for its reduction and reproduction on the printed page. For example, the hatching patterns in the legend, as well as in certain states, have coalesced to such an extent that they are not clearly distinguishable.[28]

Figure 6-12 may be succinctly characterized as anomalous since it deviates so noticeably from generally accepted principles and standards both with respect to chart design as well as with respect to the appropriateness of the type of chart selected. First, the hatching patterns are crude and amateurish. Second, the quality of the drafting is mediocre. Third, the legend is illogical and confusing. Fourth, the choropleth map technique is unsatisfctory for depicting whole numbers.[29] Fifth, because of its design and other incongruities not only is it the wrong type of chart for the data it is supposed to illustrate, it is most ineffective as a vehicle of visual communication. It would be considerably more appropriate and efficient to portray the geographical distribution of whole

Figure 6-12. An unusual as well as awkward application of the choroplethic technique. This map attempts to portray whole numbers as represented by eight class intervals. The cross-hatching scheme is crude and inconsistent. Both the basic design and mechanical features of this chart would be rated as inferior. (From Department of Justice, Immigration and Naturalization Service, 1976 Annual Report, Washington, D.C.: Government Printing Office, 1978, p. 26.)

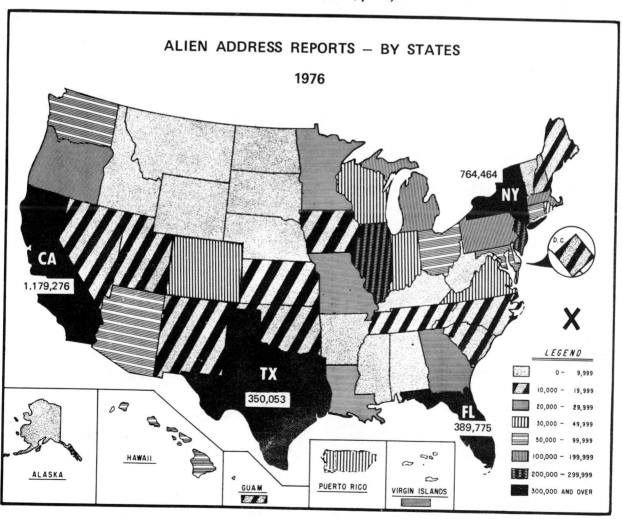

GEOGRAPHICAL DISTRIBUTION OF CRIMINAL HOMICIDE

UNITED STATES: 1958-1962

EXPLANATORY NOTE:

THE SYMBOLS ON THE MAP ARE
THREE-DIMENSIONAL. THE NUMBER
OF CASES REPRESENTED BY EACH SYMBOL
IS PROPORTIONAL TO ITS VOLUME.

NUMBER OF CASES

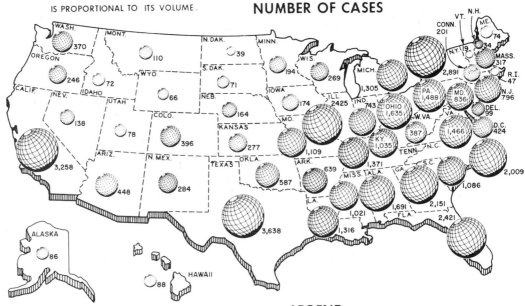

INSETS FOR ALASKA AND HAWAII OUT-OF-SCALE

LEGEND
RATE PER 100,000 OF POPULATION

UNDER 2.0	4.0-5.9	8.0-9.9
2.0-3.9	6.0-7.9	10.0-11.9

MEAN RATE PER 100,000 OF POPULATION

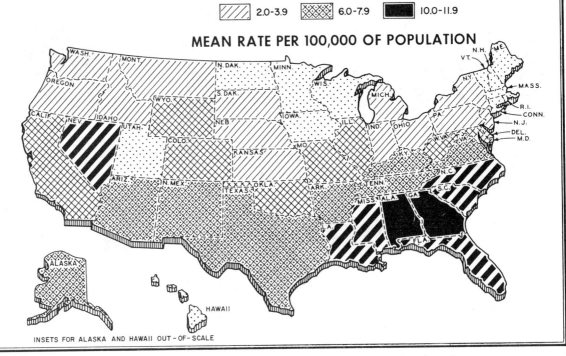

INSETS FOR ALASKA AND HAWAII OUT-OF-SCALE

Figure 6-13. The two maps in this chart complement each other. One map portrays the actual number of cases by means of three-dimensional point symbols; the other map, relative numbers by means of graded cross-hatching. The same techniques could have been used for the type of data represented in Figure 6-12. (From Calvin F. Schmid and Stanton E. Schmid, Crime in the State of Washington, Olympia: Law and Justice Planning Office, Planning and Community Affairs Agency, 1972, p. 74.)

numbers by one-, two-, or three-dimensional symbols. Of course, if a series of rates or some other kind of index were to be mapped, the choropleth technique would be satisfactory. For example, this could be readily accomplished with the data on aliens, by computing rates based on the number of registered aliens in relation to the total population in each state.

Figure 6-13 illustrates how two series of data that are statistically similar to those that might be derived

[29] Occasionally such maps may be found in print, but I believe that the use of the choroplethic technique for this purpose is inappropriate. Graduated point symbols or bars and columns would be more satisfactory in depicting absolute numbers.

from the data in Figure 6-12 can be shown graphically. The upper map portrays absolute data—the number of homicidal deaths occurring in each state during the five-year period 1958–1962, while the lower map portrays rates—the number of homicides per 100,000 of population. In the upper map, the size of each spherical symbol is commensurate with the number of cases. The name of each state as well as the number of homicides is clearly shown. The lower map depicts the mean rate per 100,000 of population according to the choropleth map technique. There are six hatching patterns ranging from light stippling to black, representing graded values with uniform intervals of 1.0.

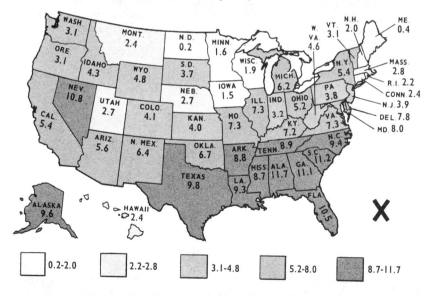

Figure 6-14. The shading on this map was printed by means of a tint screen process. In spite of the small number of shaded categories and their simplicity, the results are unsatisfactory. It is difficult to discriminate visually between certain categories, particularly the third and fourth. (From Staff Report, Donald J. Mulvihill and Melvin M. Tumin, with Lynn A. Curtis, Crimes of Violence, Vol. 11, National Commission on the Causes and Prevention of Violence, Washington, D.C.: Government Printing Office, 1969, p. 72.)

Source: UCR, 1967.

Variation in reported criminal homicide and nonnegligent manslaughter offense rates, by state, 1967, [rates per 100,000 population].

TOTAL FERTILITY RATE BY STATE, 1975

Figure 6-15. *An outline map with data superimposed on its 50 areal units. As it now stands, it serves much the same purpose as a statistical table. In order to make this map an effective medium of visual communication, it could be converted into a choropleth map by organizing the data into class intervals and developing a shading or hatching scheme. (From Charles P. Kaplan and Thomas Van Valey,* Census '80: Continuing the Factfinder Tradition, *United States Bureau of the Census, Washington, D.C.: Government Printing Office, 1980, p. 220.)*

Figure 6-16. *A choropleth map with eight hatching patterns plus an unshaded category. See text for additional comments. (From Calvin F. Schmid and Stanton E. Schmid,* Crime in the State of Washington, *Olympia: Law and Justice Planning Office, Planning and Community Affairs Agency, 1972, p. 243.)*

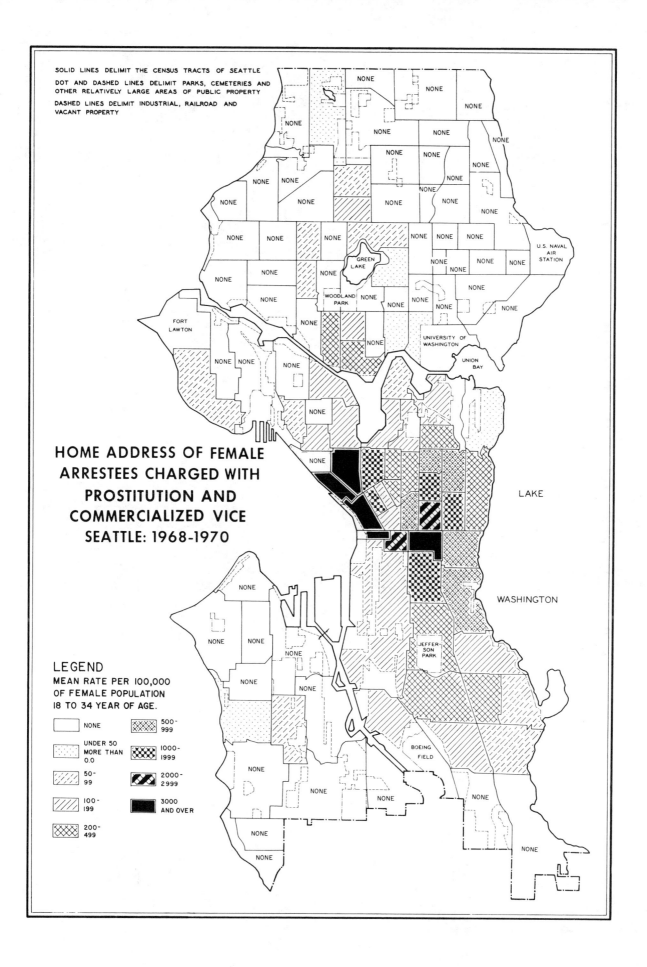

SOLID LINES DELIMIT THE CENSUS TRACTS OF SEATTLE

DOT AND DASHED LINES DELIMIT PARKS, CEMETERIES AND
OTHER RELATIVELY LARGE AREAS OF PUBLIC PROPERTY

DASHED LINES DELIMIT INDUSTRIAL, RAILROAD AND
VACANT PROPERTY

HOME ADDRESS OF FEMALE
ARRESTEES CHARGED WITH
PROSTITUTION AND
COMMERCIALIZED VICE
SEATTLE: 1968-1970

LEGEND
MEAN RATE PER 100,000
OF FEMALE POPULATION
18 TO 34 YEAR OF AGE.

NONE	500-999
UNDER 50 MORE THAN 0.0	1000-1999
50-99	2000-2999
100-199	3000 AND OVER
200-499	

CHANGES IN PUBLIC SCHOOL ENROLLMENT, GRADES 1-12
STATE OF WASHINGTON: 1960 TO 1966

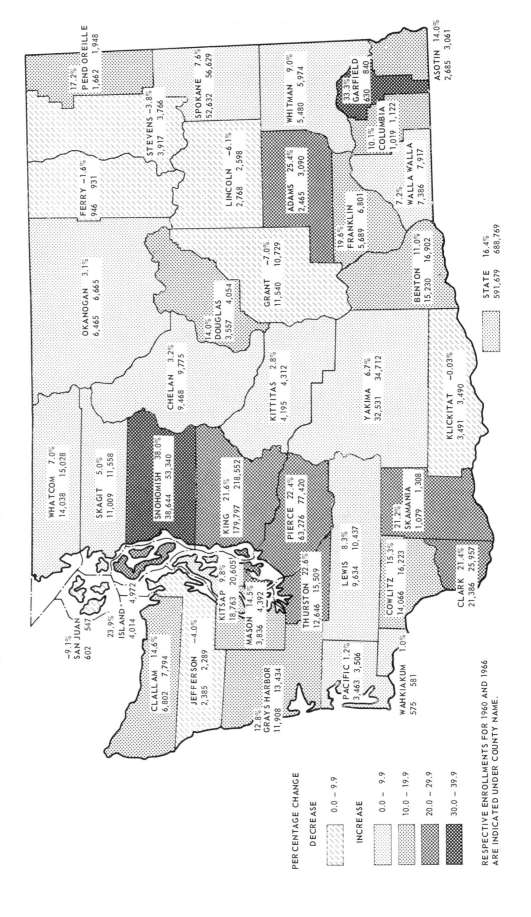

PEND OREILLE 17.2%
1,662   1,948

STEVENS -3.8%
3,917   3,766

SPOKANE 7.6%
52,632   56,629

WHITMAN 9.0%
5,480   5,974

GARFIELD 33.3%
630   840

ASOTIN 14.0%
2,685   3,061

FERRY -1.6%
946   931

LINCOLN -6.1%
2,768   2,598

ADAMS 25.4%
2,465   3,090

COLUMBIA 10.1%
1,019   1,122

OKANOGAN 3.1%
6,465   6,665

GRANT -7.0%
11,540   10,729

FRANKLIN 19.6%
5,689   6,801

WALLA WALLA 7.2%
7,386   7,917

DOUGLAS 14.0%
3,557   4,054

BENTON 11.0%
15,230   16,902

CHELAN 3.2%
9,468   9,775

KITTITAS 2.8%
4,195   4,312

YAKIMA 6.7%
32,531   34,712

KLICKITAT -0.03%
3,491   3,490

WHATCOM 7.0%
14,038   15,028

SKAGIT 5.0%
11,009   11,558

SNOHOMISH 38.0%
38,644   53,340

KING 21.6%
179,797   218,552

PIERCE 22.4%
63,276   77,420

SAN JUAN -9.1%
602   547

ISLAND 23.9%
4,014   4,972

KITSAP 9.8%
18,763   20,605

MASON 14.5%
3,836   4,392

THURSTON 22.6%
12,646   15,509

SKAMANIA 21.2%
1,079   1,308

CLALLAM 14.6%
6,802   7,794

JEFFERSON -4.0%
2,385   2,289

LEWIS 8.3%
9,634   10,437

COWLITZ 15.3%
14,066   16,223

CLARK 21.4%
21,386   25,957

GRAYS HARBOR 12.8%
11,908   13,434

PACIFIC 1.2%
3,463   3,506

WAHKIAKUM 1.0%
575   581

STATE 16.4%
591,679   688,769

PERCENTAGE CHANGE

DECREASE

0.0 – 9.9

INCREASE

0.0 – 9.9

10.0 – 19.9

20.0 – 29.9

30.0 – 39.9

RESPECTIVE ENROLLMENTS FOR 1960 AND 1966
ARE INDICATED UNDER COUNTY NAME.

Figure 6-14 is an example of a choropleth map in which the shading is printed by a tint screen process. The quality of maps reproduced in this manner can be affected by the characteristics and type of screen, the printing press, and grade of paper. Figure 6-14 fails to provide the readily discriminable tonal qualities so essential to an effective choropleth map. The map is relatively simple, with only five class intervals. Moreover, one interval is white. Even with only four remaining class intervals, it is difficult to differentiate visually certain categories, particularly the third and fourth. A different form of shading technique and less reduction could provide a map that would be more readily understood as well as being a more effective medium of communication. In this connection, it is suggested that a comparison be made between Figure 6-14 and the choropleth map in Figure 6-13. Both maps show the incidence of criminal homicide in the United States, but the clarity and discriminability of the hatching scheme in Figure 6-13, along with other features of the choropleth map, make for a better type of graphic presentation.

Not infrequently, one finds outline maps with certain statistical data superimposed. Figure 6-15 is an example of such a map. The data represent total fertility rates for each of the 50 states in 1975.[30] The purpose of constructing maps of this type may seem puzzling since they convey the same information as a statistical table, though sometimes not as effectively. Although maps of this kind emphasize the spatial character of the data, they lack the essential qualities of an effective medium of visual communication.

An effective statistical map is a visual composition consisting of a complex of interdependent design elements and tangible and intangible qualities. Tangible or structural elements may include symbols of varying size and shape, lines of distinct widths, lengths, and weights, variety of shading patterns with different textures and intensity, lettering of variable weight and size, and varying amounts of blackness and lightness. Less tangible elements and qualities include harmony, balance, and proportion of design, clarity, reliability, unity, contrast, perceptibility, and interpretability.

Besides being perceived merely as a source of statistical data with a spatial referent, Figure 6-15 may also be perceived as an uncompleted choropleth map. In fact, as it now stands, a base map with areal units (states) in outline form and detailed statistical data (fertility rates) allocated for each unit, it embodies a typical stage in the preparation of a choropleth map. In order to transform this uncompleted "work map" into a finished map, three additional steps are required: (1) The data must be organized into class intervals; (2) An appropriate shading system must be developed; and (3) The shading patterns, legend, and miscellaneous or supplementary details must be superimposed on the map.

EXAMPLES OF ACCEPTABLE CHOROPLETH MAPS

Figure 6-16 is a well-designed map with eight different hatching patterns plus a blanked or unshaded category. The unshaded category represents a specific, significant, and essential classification, that is, census tracts in which no arrestees were reported along with relatively large vacant areas in which no people, or at least very few people, reside. The variant areas include parks, cemeteries, air fields, and industrial property. In the case of census tracts with no arrestees, identification is clearly indicated by the word "none." Sociologically and ecologically, of course, the "none" and "vacant" categories are just as significant as areas with designated rates. The rates are for a two-year average and are based on the female population between the ages of 18 and 34, the age-group in which the overwhelming proportion of violations for this offense occurs. The hatching was done in pen and ink directly on the map.

---

[30] "The total fertility rate is the number of births that 1,000 women would have in their lifetime if, at each year of age, they experienced the birth rates occurring in the specified calendar year. It should be stressed that the total fertility rate is an annual (or period) measure of fertility, even though it is expressed as a hypothetical lifetime (or cohort) measure." Charles P. Kaplan and Thomas Van Valey, *Census '80: Continuing the Factfinder Tradition,* United States Bureau of the Census, Washington, D.C.: Government Printing Office, 1980, A. 221.

Figure 6-17. A choropleth map portraying rates of change in public school enrollment. A clear and simple hatching scheme plus supplementary data placed inconspicuously on the map provide a readily interpretable source of information, both in terms of visual and statistical communication. (From Calvin F. Schmid and Vincent A. Miller, Enrollment Forecasts, State of Washington: 1967 to 1975, *Seattle: Washington State Census Board, 1967, p. 18.)*

Figure 6-17 shows the increase and decrease in public school enrollment (grades 1–12) from 1960 to 1966 and for the 39 counties in the state of Washington. There are six class intervals ranging from a maximum decrease of 9.9 percent to a maximum increase of 39.9 percent. The shading patterns were superimposed on the base map from commercially preprinted, self-adhesive sheets. It also will be observed that the name of each county is clearly indicated along with the number of children enrolled both in 1960 and 1966 as well as the corresponding percentage change.

TWO- AND THREE-DIMENSIONAL GRAD-
uated point symbols, including circles, squares, cubes,
spheres, wedges, and triangles, represent a long-
standing and persistent issue in statistical graphics.
Among these symbols, the circle is the one most frequently used and discussed, as well as the one most
often subjected to experimental testing and evaluation. For many decades, caveats against the use of
graduated point symbols have appeared in the literature on graphic presentation. For example, in 1914,
Willard C. Brinton's *Graphic Methods* contains the
following admonition: "Circles compared on an area
basis mislead the reader by causing him to underestimate the ratios. Circles of different size should never
be compared."[1]

Nevertheless, in spite of the many warnings and
criticisms of this kind, two- and three-dimensional
graduated point symbols have continued to be utilized, especially in the construction of certain types of
maps. In fact, cartographers and other graphic specialists seem firmly committed to the uses of graduated point symbols.

Moreover, for example, it is significant to note
that

> The graduated circle is one of the oldest of the
> quantitative point symbols used for statistical rep-
> resentation. Near the beginning of the nineteenth
> century, it was used in graphs that illustrated the
> then new census materials, and its first appearance
> on maps was in the 1830's. Since that time, it has
> been near the top of any list of quantitative point
> symbols in the frequency of its use; its ease of con-
> struction makes it likely that it will continue to be
> widely utilized.[2]

AN OVERVIEW OF
PSYCHOPHYSICAL RESEARCH:
SOME FURTHER COMMENTS

It has been common practice for graphic specialists to
select symbols on the basis of tradition, trial and
error, and personal experience and impressions. As
indicated previously (Chapter 1), some 50 years ago a

CHAPTER SEVEN

TWO- AND THREE-DIMENSIONAL GRADUATED POINT SYMBOLS

A Perennial Issue in Statistical Graphics

few statisticians attempted to analyze by means of
tests and experimentation various symbols and
graphic forms along with certain psychophysical factors and processes. They were interested in learning
more precisely how people interpret what they see in
a chart, which, presumably, would provide more objective and reliable criteria for chart design and construction.[3]

Subsequently, psychologists, psychophysicists, ed-
ucationists, and, later, cartographers joined in this effort. During the 1960's and 1970's, cartographers
have virtually dominated this type of research. The

---

[1] Willard C. Brinton, *Graphic Methods for Presenting Facts,*
New York: The Engineering Magazine Company, 1914, p.
37.

[2] Arthur Robinson, Randall Sale, and Joel Morrison, *Elements of Cartography,* New York: John Wiley & Sons,
1978, p. 207.

---

[3] For example, a few of these early studies include Walter
Crosby Eells, "The Relative Merits of Circles and Bars for
Representing Component Parts," *Journal of the American
Statistical Association,* **21** (1926), 119–132; R. Von Huhn,
"Further Studies in Graphic Circles and Bars: A Discus-
sion of Eells' Experiment," *Journal of the American Statis-
tical Association,* **22** (1927), 31–36; Frederick E. Croxton,
"Further Studies in the Graphic Use of Circles and Bars:
Some Additional Data," **22** (1927), 36–39; Frederick E.
Croxton and R. E. Stryker, "Bar Charts Versus Circle Dia-
grams," *Journal of the American Statistical Association,* **22**
(1927), 473–482; Frederick E. Croxton and Harold Stein,
"Graphic Comparisons by Bars, Squares, Circles, and
Cubes," *Journal of the American Statistical Association,* **27**
(1932), 54–60.

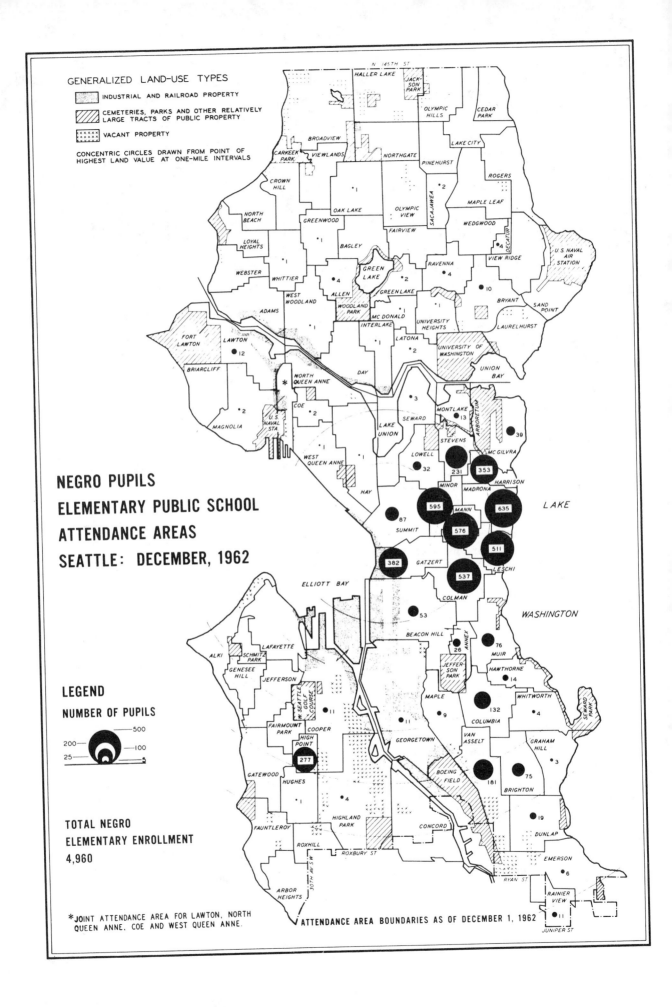

GENERALIZED LAND-USE TYPES

INDUSTRIAL AND RAILROAD PROPERTY

CEMETERIES, PARKS AND OTHER RELATIVELY LARGE TRACTS OF PUBLIC PROPERTY

VACANT PROPERTY

CONCENTRIC CIRCLES DRAWN FROM POINT OF HIGHEST LAND VALUE AT ONE-MILE INTERVALS

NEGRO PUPILS
ELEMENTARY PUBLIC SCHOOL
ATTENDANCE AREAS
SEATTLE: DECEMBER, 1962

LEGEND

NUMBER OF PUPILS

500
200 — 100
25 — 5

TOTAL NEGRO ELEMENTARY ENROLLMENT
4,960

*JOINT ATTENDANCE AREA FOR LAWTON, NORTH QUEEN ANNE, COE AND WEST QUEEN ANNE.

ATTENDANCE AREA BOUNDARIES AS OF DECEMBER 1, 1962

rationale behind research of this kind as expressed by two cartographers is succinctly summarized by the following quotations:

A map, whatever its purpose, scale or design, is a visual communicative device and is, therefore, subject to the psychophysical limitations imposed by human perception. Understanding how a myriad of different cartographic symbols are perceived presents cartographers with an enormous task. Obviously, no single study can encompass this entire problem. A sufficient number of investigators into micro-segments of the larger problem could eventually lead to an improved understanding of the psychophysical complexities of cartographic communication.[4]

Principles of map design must inevitably be based upon precise knowledge of relationships between user capabilities and symbol systems which underlie cartographic technique. The subject area has proven to be extremely complex and difficult to study. Effective strategies appear, at the least, to entail a hybridization of such disciplines as cost/benefit analysis, general systems theory, human factors, psychophysics, quantitative science, linguistics, communication theory, data processing, and the philosophy of science. Clearly the task of improving map design is a formidable one, yet steps are slowly being made, usually by means of methodologies perfected to some extent from one or more of the above fields.[5]

PSYCHOPHYSICAL RESEARCH AND GRADUATED POINT SYMBOLS

Although these statements are expressed in the context of thematic maps, they are, of course, applicable to all forms of statistical graphics. Since our primary interest in this chapter revolves around the problem of two- and three-dimensional graduated point symbols, let us review some of the relevant literature pertaining to this problem in order to ascertain what practical contributions have been derived as guides in the construction of more valid and effective statistical charts.

In order to assess the validity and reliability, as well as the nature and applicability of research of this kind, it is essential to possess substantial insight into the methodology and data that have been utilized. Generally, classical psychophysical techniques have been employed in this type of research with particular emphasis on discovering perceptual correlates of spatial dimensions.

Although the "reading" or "decoding" of statistical charts, especially the more elaborate ones, represents a complicated process involving a number of stages, the direction of most of the research has been limited to analyses of individual symbols. It will be found that a large number of these studies have been subjected to criticism because of methodological and other deficiencies. For example, most often the studies of symbols have not been made within the total context of statistical charts, but rather with isolated, disparate test cards. The obvious implication of such a procedure is that test cards provide an artificial and unrealisitic evaluative setting for understanding statistical charts in their totality. Also, in some instances, charts used as basic experimental models have been so poorly structured as to vitiate the data as well as the resultant conclusions. Again, there are studies that have suffered from other methodological deficiencies, including improper or inadequate statistical analysis.[6]

---

[4] Paul V. Crawford, "The Perception of Graduated Squares as Cartographic Symbols," *Cartographic Journal,* **10** (1973), 85–94.

[5] Phillip C. Muehrcke, *A Psychophysical Investigation of Cartographic Techniques* (Technical Report), Environmental Science Division, U.S. Army Research Office (Durham) Contract No. DAHCO4-74-G-0019, Durham, North Carolina, August 1974, Appendix A, p. A1.

[6] For more detailed discussions of issues and problems of this kind see, for example, George F. McCleary, Jr., "Beyond Simple Psychophysics: Approaches to the Understanding of Map Perception," *Proceedings of the American Congress on Surveying and Mapping* (1970), 189–209; C. Board and R. M. Taylor, "Perception and Maps: Human Factors in Map Design and Interpretation," *Transactions, Institute of British Geographers,* New Series, **2** (1977), 19–36; Howard Wainer and Albert D. Biderman, "Some Methodological Comments on Evaluating Maps," *The Cartographic Journal,* **14** (1977), 109–114.

Figure 7-1. Map with two-dimensional proportional symbols representing values for entire areas. Portrays the number of black children in each district attending public elementary school. (From Calvin F. Schmid and Wayne W. McVey, Growth and Distribution of Minority Races in Seattle, Washington, Seattle: Seattle Public Schools, 1964, p. 30.)

GRADUATED CIRCLES

With respect to the various graduated symbols used in statistical graphics, James J. Flannery lists six "distinct advantages in favor of the graduated circle": (1) ease in converting basic quantitative data into symbolic form, (2) short time required to draw symbols, (3) efficient use of space, (4) circles are more aesthetic than squares and other symbols, (5) accurate portrayal of distribution patterns, and (6) relatively greater effectiveness visually when used as pie charts.[7] In addition, a seventh advantage might be included: The circle is not influenced by the size and shape of the area it represents ("The curse of the choropleth").[8]

In a survey representing a sample of 1040 students in five colleges and universities, Flannery found a persistent underestimation of the perceived magnitude of circles graduated on an areal size basis. Other studies have revealed similar results, particularly when larger circles were compared to smaller ones.

SQUARE-ROOT METHOD OF CONSTRUCTING PROPORTIONAL CIRCLES

As previously indicated, circles are two-dimensional and represent areas. The area of a circle is equal to its radius squared times the constant pi (π); that is, $A = \pi r^2$. Mathematically, the area of the circle is a function of the square of the diameter, the area increasing directly as the square of the diameter. In laying out circles as symbols for comparison of sizes, the diameters of the symbols should be proportional to the square root of the sizes to which they correspond.

Accordingly, the first step in determining the sizes of a series of circular symbols is to extract the square roots of the figures they are to represent. The circles are drawn with radii or diameters proportional to the derived square roots. It is important to construct circles compatible with the general layout and dimensions of the chart.[9] This scaling procedure reflects the values of the original data most accurately and is known as the "square-root method."

SCALING TECHNIQUES FOR MINIMIZING PERCEPTUAL DISCREPANCIES

However, in light of the fairly consistent tendency to misperceive the "real" areas of graduated circles, especially when larger symbols are compared to smaller ones, various measures have been recommended to minimize perceptual discrepancies. Instead of scaling proportional circles precisely on the basis of the data which they represent, they are scaled according to some constant that would progressively expand the circles. By this method, the size of the resultant symbols would reflect a degree of compensation for underestimation. This technique has been referred to as the "psychological scaling method." In a sense, it is a form of "visual trickery" designed to facilitate a more reliable interpretation of size in the "eye–brain" system.[10]

The crucial problem, of course, in gauging the effectiveness of this method is to determine the validity and reliability of the constant that is used.[11] Several studies have shown that with respect to circles, squares, and triangles, the judged area (perceived size) as a power function of physical area (actual size) has ranged very widely. For circles, most of the studies have reported factors ranging from .70 to .95.

Of course, the various exponents derived from these studies are not comparable because the studies generally differ in (1) research objectives; (2) test design, test procedure, and so on; (3) sample size and composition; (4) "averages" and "standards" selected in the analyses; and (5) statistical methodology.[12]

ROLE OF THE LEGEND IN MINIMIZING PERCEPTUAL ERROR

Since psychological techniques alone do not compensate fully for the tendency to underestimate the size of

[7] James John Flannery, "The Relative Effectiveness of Some Common Graduated Point Symbols in the Presentation of Quantitative Data," *The Canadian Cartographer,* **8** (1971), 96–109.

[8] Hans-Joachim Meihoefer, "The Utility of the Circle as an Effective Cartographic Symbol," *The Canadian Cartographer,* **6** (1969), 105–117.

[9] For a more detailed discussion of the procedure for constructing graduated two- and three-dimensional symbols, see Calvin F. Schmid and Stanton E. Schmid, *Handbook of Graphic Presentation,* New York: John Wiley & Sons, 1979, pp. 194–196.

[10] A more detailed discussion of this procedure will be found in Arthur Robinson, Randall Sale, and Joel Morrison, *Elements of Cartography,* New York: John Wiley & Sons, 1978, pp. 208–209.

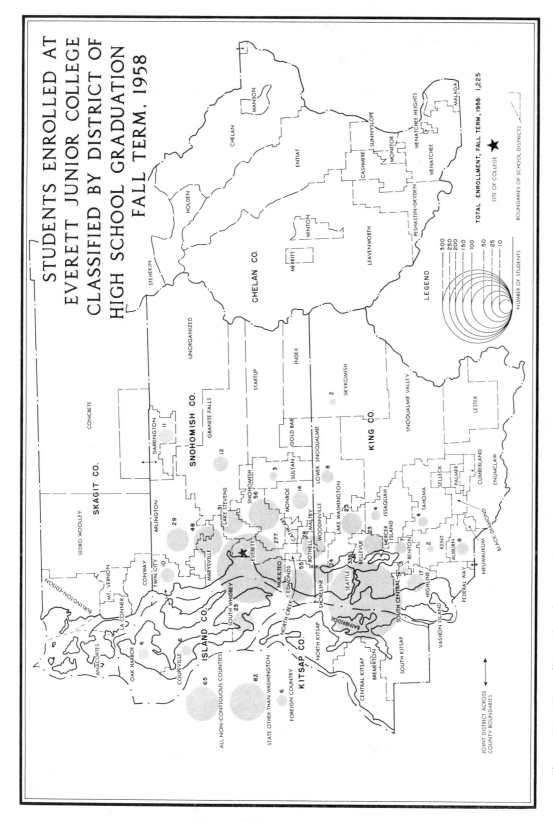

Figure 7-2. **Two-dimensional proportional symbols shown in gray tone. Black symbols as portrayed in Figures 7-1 and 7-3 are frequently misperceived as smaller than open (white) or gray-tone symbols. This map also indicates values for entire areas. The data represent the sources of the student body of Everett Junior College classified according to the high schools from which the students were graduated. (From** *Calvin F. Schmid and Vincent A. Miller, Population Trends, Educational Change,* **Seattle: Washington State Census Board, 1960, p. 132.)**

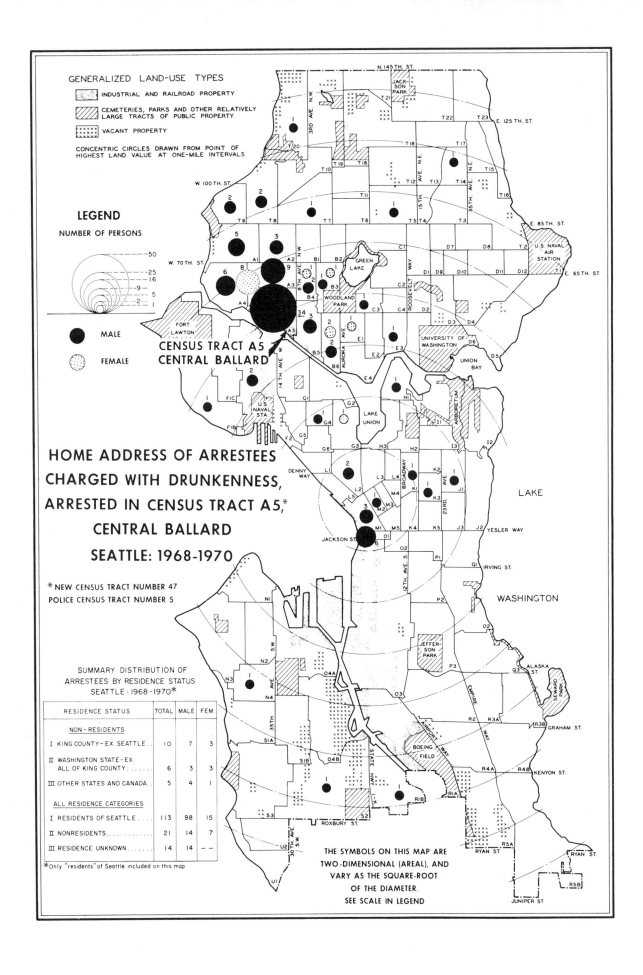

GENERALIZED LAND-USE TYPES

INDUSTRIAL AND RAILROAD PROPERTY

CEMETERIES, PARKS AND OTHER RELATIVELY LARGE TRACTS OF PUBLIC PROPERTY

VACANT PROPERTY

CONCENTRIC CIRCLES DRAWN FROM POINT OF HIGHEST LAND VALUE AT ONE-MILE INTERVALS

LEGEND

NUMBER OF PERSONS

— 50
— 25
— 16
— 9
— 5
— 2
— 1

● MALE

○ FEMALE

CENSUS TRACT A5 CENTRAL BALLARD

HOME ADDRESS OF ARRESTEES CHARGED WITH DRUNKENNESS, ARRESTED IN CENSUS TRACT A5, CENTRAL BALLARD SEATTLE: 1968-1970

* NEW CENSUS TRACT NUMBER 47 POLICE CENSUS TRACT NUMBER 5

SUMMARY DISTRIBUTION OF ARRESTEES BY RESIDENCE STATUS SEATTLE : 1968-1970*

RESIDENCE STATUS	TOTAL	MALE	FEM.
NON-RESIDENTS			
I KING COUNTY-EX. SEATTLE..	10	7	3
II WASHINGTON STATE-EX. ALL OF KING COUNTY......	6	3	3
III OTHER STATES AND CANADA..	5	4	1
ALL RESIDENCE CATEGORIES			
I RESIDENTS OF SEATTLE....	113	98	15
II NONRESIDENTS............	21	14	7
III RESIDENCE UNKNOWN......	14	14	—

*Only "residents" of Seattle included on this map

THE SYMBOLS ON THIS MAP ARE TWO-DIMENSIONAL (AREAL), AND VARY AS THE SQUARE-ROOT OF THE DIAMETER. SEE SCALE IN LEGEND

circles, special attention has been given to other scaling adjustments and to chart legends as ancillary or supplementary measures.

Hans-Joachim Meihoefer suggests that the problem of the map reader being unable to discern small variations in circle size can be partially rectified by changing the method of presenting the sizes of circles in the map legend. Instead of using the traditionally popular method of showing graduated circles in the legend so that circle size is interpreted in terms of the value of each individual figure, the circles are range-graded, that is, grouped by class intervals. The sizes of circles on a map are gauged by symbols in an accompanying legend representing a range-graded series, the size of each symbol being determined by the midvalue of each class interval. For example, in portraying several cities of varying sizes on a map, there would be one symbol of a certain diameter for cities with populations say, that range from 25,000 to 49,-999, another symbol for cities of 50,000 to 99,999, and so on. Using this procedure, every circle on the map would be represented by a circle of corresponding size in the legend.[13]

In designing a map legend, it also has been recommended that certain selected referents or "anchor stimuli" be included. These referents are in the form of several symbols that represent those portrayed on the map. The symbols in the legend serve as "anchor stimuli" by assisting the reader in interpreting more accurately the values of circles on the map.[14]

In constructing graduated or proportional point symbol maps, whether they represent circles or spheres, the present writer includes a well-planned legend, except where the data are very simple and straightforward. In addition, however, a precise referent figure is always included for every symbol on the map, no matter how simple the map may be. Meihoefer feels that this additional information can help minimize peceptual and interpretational error.

In case there is a misinterpretation or uncertainty in the visual perception process, the figures provide a ready and convenient check on the value of every symbol. This practice can be characterized as "useful redundancy," a design principle that enhances the clarity and comprehension of a chart by repetitive emphasis.

Perhaps the only potential criticism of this practice is that it could create clutter, but maps constructed in this manner do not seem to exhibit this weakness. Figures 7-1 to 7-4 and Figure 7-8 are typi-

---

[11] In this connection, see James J. Flannery, *The Graduated Circle: A Description, Analysis and Evaluation of a Quantitative Map Symbol* (unpublished doctoral dissertation), University of Wisconsin, 1956. A few additional references of psychological and cartographic papers relating to area and volume estimation are as follows: Gosta Ekman and Kenneth Junge, "Psychological Relations in the Perception of Length, Area and Volume," *Scandinavian Journal of Psychology,* **2** (1961), 1–10; Gosta Ekman, Ralf Lindman, and W. Williams-Olsson, "A Psychological Study of Cartographic Symbols," *Perceptual and Motor Skills,* **13** (1961), 355–368; S. S. Stevens, "On the Psychophysical Law," *The Psychological Review,* **64** (1957), 153–181; Paul V. Crawford, "Perception of Grey-tone Symbols," *Annals, Association of American Geographers,* **61** (1971), 721–735; Borden D. Dent, "Communication Aspects of Value-by-Area Cartograms," *The American Cartographer,* **2** (1975), 154–168; R. L. Williams, *Statistical Symbols for Maps; Their Design and Relative Values,* Office of Naval Research Paper, New Haven: Yale University, 1956; M. Teghtsoonian, "The Judgement of Size," *The American Journal of Psychology,* **78** (1965), 392–402; William S. Cleveland, Charles S. Harris, and Robert McGill, "Circle Sizes for Thematic Maps," *Urban, Regional, and State Government Applications of Computer Mapping,* Cambridge, Mass.: Harvard Library of Computer Graphics, 1980, Vol. 11, pp. 40–47; Kang-tsung Chang, "Circle Size Judgment and Map Design," *The American Cartographer,* **7** (1980), 115–162; Kang-tsung Chang, "Visual Estimation of Graduated Circles," *The Canadian Cartographer,* **14** (1977), 130–138.

[12] Paul V. Crawford, "The Perception of Graduated Squares as Cartographic Symbols," *Cartographic Journal,* **10** (1973), 85–94.

[13] Hans-Joachim Meihoefer, "The Utility of the Circle as an Effective Cartographic Symbol," *The Canadian Cartographer,* **6** (1969), 105–117; Hans-Joachim Meihoefer, "The Visual Perception of the Circle in Thematic Maps/Experimental Results," *The Canadian Cartographer,* **10** (1973), 63–84.

[14] For more detailed discussions, see Carleton W. Cox, "Anchor Effects and the Estimation of Graduated Circles and Squares," *The American Cartographer,* **3** (1976), 65–74.

Figure 7-3. Chart with two sets of proportional areal symbols. It depicts the residence by census tract of persons arrested for drunkenness in an outlying business district of Seattle (Ballard). The arrestees are differentiated by sex: Black symbols represent males and stippled symbols represent females. (From Calvin F. Schmid and Stanton E. Schmid, Crime in the State of Washington, Olympia: Law and Justice Planning Office, Washington State Planning and Community Affairs Agency, 1972, p. 304.)

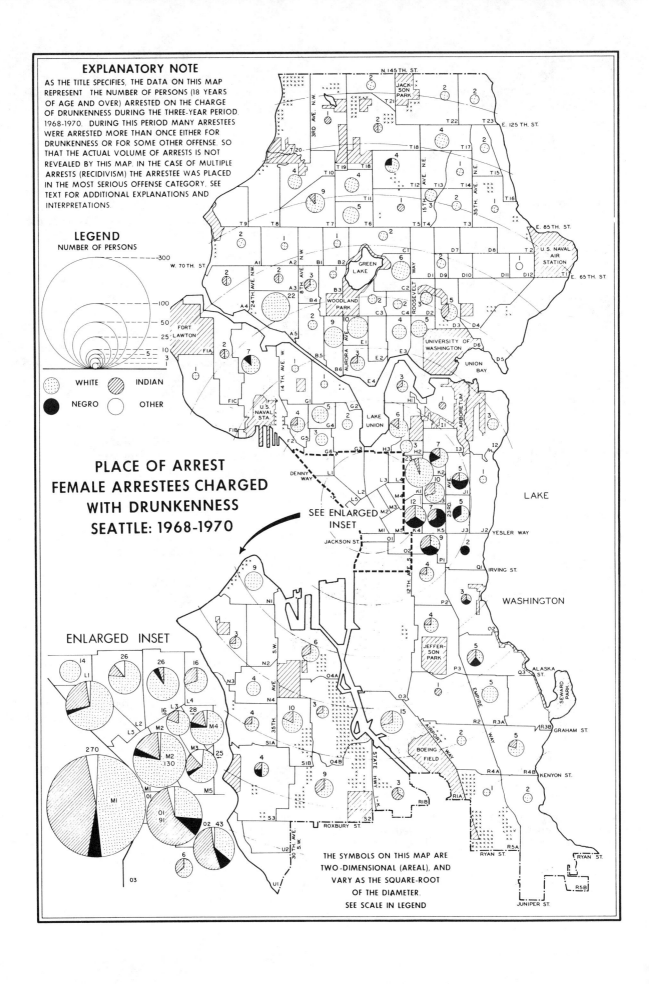

EXPLANATORY NOTE

AS THE TITLE SPECIFIES, THE DATA ON THIS MAP
REPRESENT THE NUMBER OF PERSONS (18 YEARS
OF AGE AND OVER) ARRESTED ON THE CHARGE
OF DRUNKENNESS DURING THE THREE-YEAR PERIOD,
1968-1970. DURING THIS PERIOD MANY ARRESTEES
WERE ARRESTED MORE THAN ONCE EITHER FOR
DRUNKENNESS OR FOR SOME OTHER OFFENSE, SO
THAT THE ACTUAL VOLUME OF ARRESTS IS NOT
REVEALED BY THIS MAP. IN THE CASE OF MULTIPLE
ARRESTS (RECIDIVISM) THE ARRESTEE WAS PLACED
IN THE MOST SERIOUS OFFENSE CATEGORY. SEE
TEXT FOR ADDITIONAL EXPLANATIONS AND
INTERPRETATIONS.

LEGEND
NUMBER OF PERSONS

—300
—100
—50
—25
—10
—5
—3

WHITE   INDIAN

NEGRO   OTHER

PLACE OF ARREST
FEMALE ARRESTEES CHARGED
WITH DRUNKENNESS
SEATTLE: 1968-1970

SEE ENLARGED
INSET

ENLARGED INSET

THE SYMBOLS ON THIS MAP ARE
TWO-DIMENSIONAL (AREAL), AND
VARY AS THE SQUARE-ROOT
OF THE DIAMETER.
SEE SCALE IN LEGEND

Figure 7-4. Chart with segmented proportional areal symbols. This map shows the place of arrest by census tract of women charged with being drunk. The symbols are divided into a fourfold racial classification: White, Negro, Indian, and Other. (From Calvin F. Schmid and Stanton E. Schmid, Crime in the State of Washington, Olympia: Law and Justice Planning Office, Washington State Planning and Community Affairs Agency, 1972, p. 295.)

Figure 7-5. Two-dimensional symbols portraying exports and imports for certain coastal and interior ports in the United States and Canada. Coastal ports are shown by open circles; river ports, by stippled circles. The volume of exports and imports for coastal ports are indicated by the varying sizes of triangles. See legend for cargo tonnage and for exports and imports. (From D. K. Adams and H. B. Rodgers, An Atlas of North American Affairs, London: Methuen and Company, 1979, p. 89.)

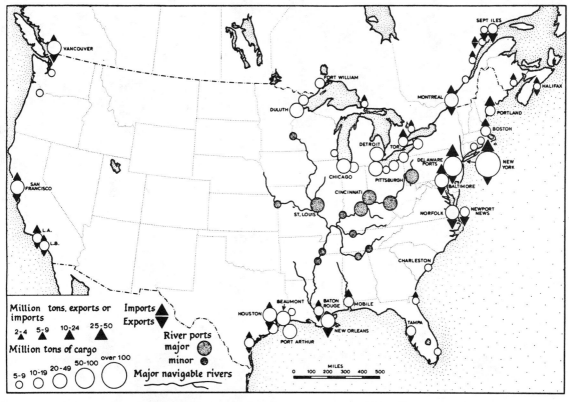

Major coastal and interior ports of North America, 1963 figures. TOR = Toronto, LA = Los Angeles, LB = Long Beach.

COLLEGE AND UNIVERSITY ENROLLMENT POTENTIALS: WASHINGTON, 1960

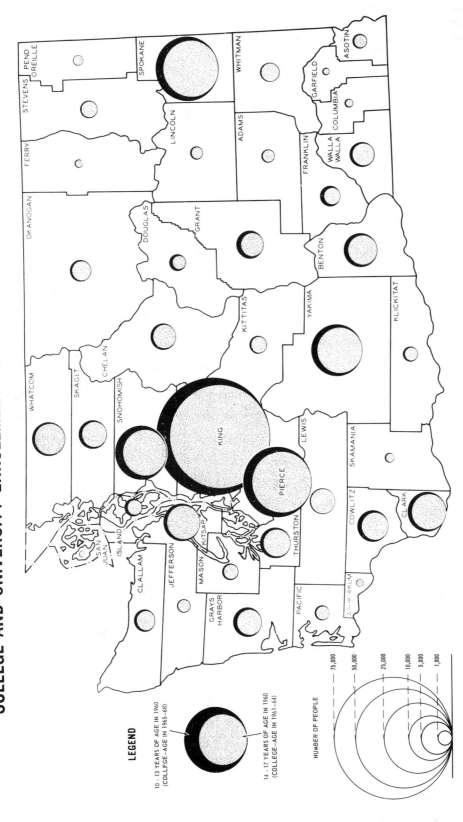

Figure 7-6. *Two-dimensional symbols designed to show change. The inner stippled circle portrays the size of the 14–17 age cohort, and the larger circle the size of the 10–13 age cohort as reported in the 1960 census. These cohorts represent the major reservoirs of college and university students for each of the 39 counties. In the academic year of 1963, the 14–17 age cohort of 1960 was the potential college and university population (18–21 years of age). Similarly, the 10–13 age cohort of 1960 represented the potential college and university population of 1968. (From Calvin F. Schmid, Vincent A. Miller, and William S. Packard, Enrollment Statistics, Colleges and Universities, State of Washington: Fall Term, 1963, Seattle: Washington State Census Board, 1964, p. 68.)*

cal examples of maps in which the precise value represented by each symbol is clearly indicated.

METHODS OF PORTRAYING GRADUATED POINT SYMBOLS IN MONOCHROMATIC CHARTS

In monochromatic charts, graduated areal symbols may be represented as simple outlines in black ink, or the entire area may be covered with black ink, cross-hatching, or a gray tone. In light of a study of the visual qualities of gray-tone symbols, Paul V. Crawford makes a strong case for the wider use of grey tones for graduated and other symbols in monochromatic charts. He states, in part, "that gray tones, when properly used, enhance the clarity, readability, and aesthetic qualities of a map ..." (see Figure 7-2).[15]

In charts where the symbols represent two or more categories of data, the symbols (customarily circles) may be divided into sectors, each sector being differentiated by a distinctive cross-hatching or shading pattern. It will be observed from Figure 7-1 that the symbols are black, and in Figure 7-2 they are gray. In Figures 7-3 and 7-5, there are two sets of symbols: One is differentiated according to sex, and the other, according to type of port. The symbols in Figure 7-4 are segmented and shaded in accordance with four racial categories.

In colored maps, the symbols are generally shown in one or more appropriate colors.

USE OF SQUARES AS GRADUATED POINT SYMBOLS

Graduated point symbols in the form of squares are also used in the construction of statistical maps but not nearly as often as circles, primarily because they are more difficult and time-consuming to make. Also, it is alleged that squares are less versatile and aes-

thetic than circles. However, Paul V. Crawford reports on the basis of one of his studies that

> The average map reader can correctly estimate the relative area of graduated squares within a range of sizes appropriate for use on small scale thematic maps. These results should not be extended to include area comparisons that are physically larger, or that have a larger range of sizes than those included in this study.[16]

PORTRAYING OVERLAPPING CLUSTERS OF GRADUATED CIRCULAR SYMBOLS

Not infrequently when graduated circular symbols are superimposed on a map, overlapping clusters may occur. This poses two problems: First, how should the symbols be drawn? Second, does the manner in which the symbols are drawn influence their interpretation, especially with respect to size? For many years, graphic specialists have used two fairly conventional patterns in delineating overlapping clusters of circular symbols. These are referred to as the "cut-out" and the "overlap" or "transparent" patterns. (See Figure 7-7.) Richard E. Groop and Daniel Cole report that "transparent" or "overlap" circles are more accurately estimated by respondents than "cut-out" circles.[17]

Some graphic specialists consider overlapping symbols undesirable and believe some alternative procedure should be used, if possible. John I. Clarke, suggests that since overlapping symbols make visual evaluation more difficult, they "can be avoided by making large-scale insets. No change in the scale of symbols is necessary."[18]

Another alternative, which will be discussed in the following section, is to use three-dimensional symbols. In comparison to two-dimensional symbols, three-dimensional symbols permit graphic representation of a much wider numerical range in considerably less space.

SPHERES AND CUBES AS GRADUATED POINT SYMBOLS

Three-dimensional graduated point symbols, usually in the form of spheres, are frequently used in statistical charts, particularly maps. Occasionally, cubic symbols may be selected for certain types of maps. In

[15] Paul V. Crawford, "Perception of Grey-Tone Symbols," *Annals of the Association of American Geographers,* **6** (1971), 721–735.

[16] Paul V. Crawford, "The Perception of Graduated Squares as Cartographic Symbols," *Cartographic Journal,* **10** (1973), 85–94.

[17] Richard E. Groop and Daniel Cole, "Overlapping Graduated Circles/Magnitude Estimation and Method of Portrayal," *The Canadian Cartographer,* **15** (1978), 114–122.

[18] John L. Clarke, "Statistical Map Reading," *Geography,* **44** (1959), 96–104.

METHODS OF SHOWING OVERLAP
OF CIRCULAR SYMBOLS

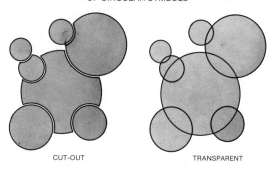

CUT-OUT   TRANSPARENT

Figure 7-7. Two conventional methods for portraying overlapping clusters of graduated circular symbols.

constructing spherical symbols, the great circle of a sphere is outlined and then shaded or marked to simulate the three-dimensional character of the symbol. (See Figures 7-8 and 7-9.) The volume of a sphere is a function of the cube root of its diameter. For example, increasing the diameter of the great circle of a sphere to three times its original length will increase the volume by 27 times its original size.[19]

Although spherical symbols possess certain limitations, perhaps more than areal symbols, they frequently serve an important function. This is especially true when the values to be presented on a map represent a relatively wide range. For example, in designing maps to portray the location, size, and growth of the cities and towns in the Puget Sound region of Washington state that ranged in population from less than 200 to well over 500,000, it was entirely impracticable to use either one- or two-dimensional symbols. A symbol scaled for a figure of over 500,000 based either on linear dimension (column or bar) or areal dimension (circle or square) would have to be overwhelming in size, and, in contrast a symbol scaled for a figure of less than 200 would have to be microscopic in size. Furthermore, the clustering of symbols in a relatively circumscribed area adds to the complexity of the problem. Figure 7-9 exemplifies how this problem was resolved by using three-dimensional symbols.

In light of psychophysical surveys of three-dimensional graduated point symbols, it has been found that there is a tendency to estimate their size on the basis of the area they cover on a chart rather than on the basis of the volume they are designed to repre-

sent. Because of perceptual deficiencies, their utility as graduated symbols is more restricted than that of two-dimensional symbols.

CONCLUSIONS

This chapter provides a summary of empirical studies along with informal opinions and recommendations relating to two- and three-dimensional graduated point symbols—their advantages, limitations, and applications. The important role that graduated symbols occupy in statistical graphics has been clearly demonstrated. In spite of this fact, some critics would abandon the use of the circle as well as other two- and three-dimensional graduated point symbols, while most cartographers and other graphic specialists consider "the circle as a valuable and useful map symbol that could be used more frequently than it is."

The pervasive significance of perceptual problems of symbolization in chart construction has been emphasized. At the same time, it has been recognized that individual symbols are only elements or components of a complete chart. In graphic communication, the end product must be a concise design in total graphic form in which all of the necessary components are integrated into a clear, meaningful visual unity that can be readily understood by the reader. Although information derived from numerous perception studies of component graphic elements has been a significant factor in the improvement of individual symbols, these studies have had little impact on fundamental chart design. To be sure, meaningful and readily interpretable symbols are essential in a well-structured chart. The message a chart is designed to communicate inheres in the total arrangement of all of its elements, including the symbols. Moreover, the perception process encompasses the entire chart, and the manner in which it is structured,

---

[19] For a description of the procedure for constructing symbols, see Calvin F. Schmid and Stanton E. Schmid, *Handbook of Graphic Presentation,* New York: John Wiley & Sons, 1979, pp. 194–196.

GEOGRAPHIC DISTRIBUTION OF NON-RESIDENT STUDENTS *
UNIVERSITY OF WASHINGTON: FALL TERM, 1965

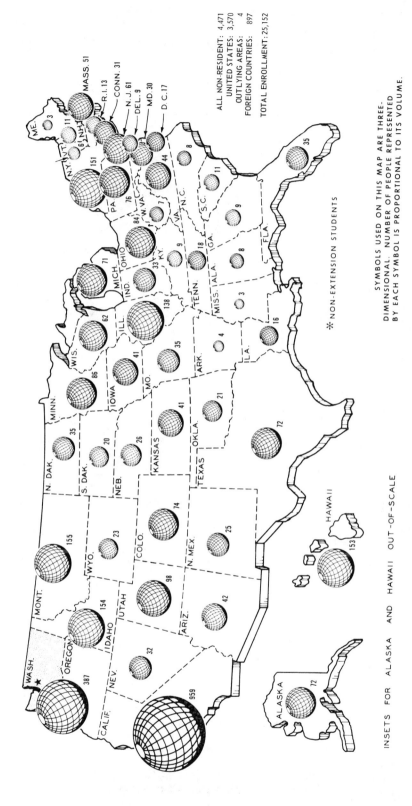

ALL NON-RESIDENT: 4,471
UNITED STATES: 3,570
OUTLYING AREAS: 4
FOREIGN COUNTRIES: 897

TOTAL ENROLLMENT: 25,152

SYMBOLS USED ON THIS MAP ARE THREE-
DIMENSIONAL. NUMBER OF PEOPLE REPRESENTED
BY EACH SYMBOL IS PROPORTIONAL TO ITS VOLUME.

*NON-EXTENSION STUDENTS

INSETS FOR ALASKA AND HAWAII OUT-OF-SCALE

*Figure 7-8. Chart with three-dimensional graduated point symbols depicting a rela-
tively simple geographic distribution. The referent for each symbol is a specified
area—one of the 50 states. (From Calvin F. Schmid, Vincent A. Miller, and William S.
Packard, Enrollment Statistics, Colleges and Universities in the State of Washington:
Fall Term, 1965, Seattle: Washington State Census Board, 1966, p. 85.)*

POPULATION CHANGE, CITIES AND TOWNS
PUGET SOUND REGION: 1920 AND 1968

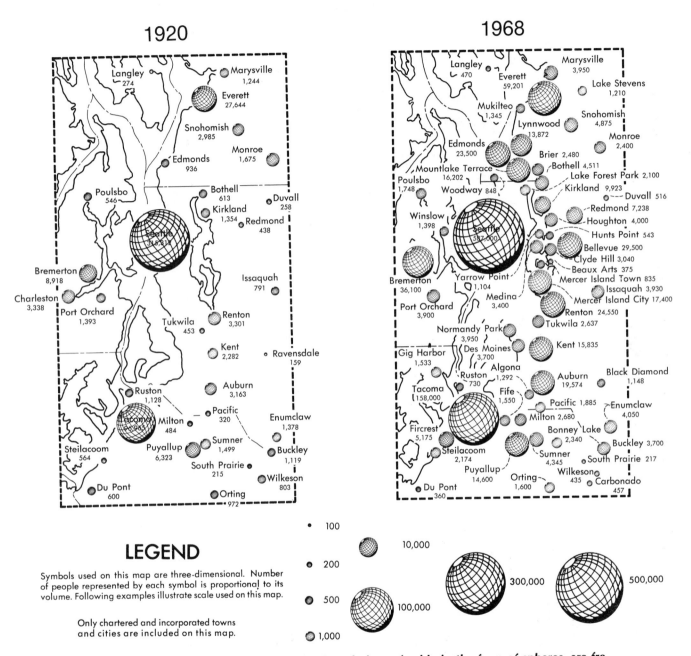

1920

1968

LEGEND

Symbols used on this map are three-dimensional. Number of people represented by each symbol is proportional to its volume. Following examples illustrate scale used on this map.

Only chartered and incorporated towns and cities are included on this map.

• 100
◦ 200
◦ 500
◎ 1,000

10,000
100,000
300,000
500,000

Figure 7-9. Three-dimensional symbols, preferably in the form of spheres, are frequently appropriate for portraying population density or the size and location of cities, towns, or other communities. This chart shows the location and size of all the chartered and incorporated cities and towns in the Puget Sound region in 1920 and in 1968. (From Calvin F. Schmid and Stanton E. Schmid, Growth of Cities and Towns, State of Washington, *Olympia: Washington State Planning and Community Affairs Agency, 1969, pp. 67–68.)*

not just the separate symbols per se.[20] Obviously, there are still many unanswered questions concerning chart design, especially in relation to graduated point symbols. Additional experimental evidence pertaining to two- and three-dimensional point symbols is needed, and, perhaps more important, greater attention should be focused on the intricate arrangements and relationships of symbols and other elements of the total chart.[21]

---

[20] In this connection, for example, George F. Jenks has prepared a careful study pertaining to new regional or pattern information derived from proportional circle maps. See George F. Jenks, "The Evaluation and Prediction of Visual Clustering in Maps Symbolized with Proportional Circles," in John C. Davis and Michael J. McCullagh, *Display and Analysis of Spatial Data,* New York: John Wiley & Sons, 1974, pp. 311–327.

[21] Compare Borden D. Dent, "Visual Organization and Thematic Map Communication," *Annals of Association of American Geographers,* **62** (1972), 79–93; Barbara Bartz, "Characterizing the Look of Maps," *American Cartographer,* **1** (1974), 63–71; Leonard Guelke, "Perception, Meaning and Cartographic Design," *The Canadian Cartographer,* **16** (1979), 61–69.

CHAPTER EIGHT

PROJECTION TECHNIQUES

Pitfalls and Problems

IN RECENT DECADES, THREE-DIMENsional charts with their attractive, picturelike features, as well as their sense of distance and solidity, have achieved an extraordinary popularity. Unfortunately, like many other potentially effective techniques in graphic presentation, they have been misused. To fully explain this state of affairs would be most difficult, but a careful examination of hundreds of three-dimensional charts seems to justify the following inferences: (1) Many of the designers of three-dimensional charts lack familiarity with the principles of projection techniques; (2) many are not conversant with the basic standards and theory of graphic presentation; and (3) some designers are inclined to emphasize novelty and cosmetic superficiality at the expense of precision, clarity, and authenticity.

Perhaps the most logical as well as pragmatic criterion in judging the acceptability of a statistical chart is its basic effectiveness as a communication device: Does it convey efficiently the message it is designed to communicate? Is it simple and readily understood? Is it attractive, forceful, and reliable? When used with understanding and discrimination, three-dimensional techniques can enhance the quality and effectiveness of a chart. However, when applied inappropriately, these techniques can be a serious nuisance and actually debase a chart.

The delineation of three-dimensional charts should not be thought of merely as the product of impressionistic artistry, but rather as the product of fairly rigorous principles and techniques of geometric projection. Moreover, not all projection techniques are equally appropriate or applicable in the design of statistical charts. For example, perspective projection should seldom be used in the preparation of statistical charts, while the various forms of axonometric and oblique projection will be found to have a much more significant and wider application.

A necessary prerequisite for designing three-dimensional charts is a thorough working knowledge of projection techniques. Projection techniques are widely used in engineering, architecture, and art to portray the shape and size of objects in accordance with certain principles and rules that govern the reproduction of a view on a flat plane. In a most simplified and abbreviated form, uniplanar (one plane) pictorial projection as generally used in statistical graphics may be classified as follows:[1]

1. Axonometric
    (a) Isometric
    (b) Dimetric
    (c) Trimetric

2. Oblique
    (a) Various oblique positions

3. Perspective
    (a) Angular
    (b) Parallel

Axonometric, oblique, and perspective projections represent one-plane pictorial representation, so that, for example, all three faces of a cube may be displayed in different forms and positions. Since an object may be placed in a countless number of positions relative to the picture plane, an infinite number of views that affect the general proportions, length of edges, and sizes of angles of the object may result. The terminology used to describe the several kinds of projection is based on the angle at which the projectors strike the plane of projection. In axonometric projection, the projectors are perpendicular to the picture plane, and the three faces of an object are visible in one view. When the principle axes make equal angles with the plane of projection, the projection is known as isometric. In dimetric projection, two of the axes make equal angles with the plane of projection, and the third is smaller or larger. In trimetric projection, each of the three axes makes a different angle with the plane of projection.

Unlike axonometric projection, the projectors in oblique projection are not perpendicular to the picture plane, but rather make an angle other than 90° with the plane of projection. There are several variations of oblique projection as determined by the angle of obliquity of the projectors with the picture plane. There is some similarity between isometric and oblique projection in that each has three axes, representing three mutually perpendicular edges upon which measurements can be made.

Of the several forms of pictorial projection, perspective is perhaps the most realistic. In perspective projections, an object is represented as normally seen by the eye of an observer located at a finite distance from the object, which is known as a station point. There are two common forms of perspective, parallel (one point) and angular (two point), as determined by the position of the picture plane with reference to the object. When the object has one face parallel to the picture plane (plane of projection) and the other faces perpendicular (90%) to it, the perspective is referred to as parallel perspective, but when the object is at an angle (other than 90°) to the picture plane, the perspective is referred to as angular perspective. In addition, there is oblique (three-point) perspective.

In applying pictorial projection techniques to the construction of statistical charts, certain inherent limitations and problems should be recognized. They may be summarized as follows: First, the resultant charts may have a distorted and unreal appearance. This fact is most common in perspective projection. Second, the actual process of preparing charts in pictorial projection may be complicated and time-consuming. Before designing a three-dimensional chart, it may be helpful to learn about the latest mechanical

devices, templates, drafting instruments, and printed forms used in the preparation of projection drawings. Also, various projection patterns can be generated by electronic computers. Third, it may be difficult to lay out precise distances and to determine exact dimensions on a projection drawing.

Although the preceding remarks are purposefully brief, the following discussion should help clarify the basic facts pertaining to projection techniques as well as provide guidance and understanding in relation to the many applications, limitations, and pitfalls of projection techniques in statistical graphics.

---

[1] For a more detailed discussion of projection techniques used in graphic presentation, see: Calvin F. Schmid and Stanton E. Schmid, *Handbook of Graphic Presentation,* New York: John Wiley & Sons, 1979, pp. 243–275. More generalized discussions of projection techniques may be found in standard treatises on technical drawing. See: Frederick E. Giesecke, et al., *Technical Drawing,* New York: Macmillan, 1974, pp. 153–202 and 497–566; Warren J. Luzadder, *Fundamentals of Engineering Drawing,* Englewood Cliffs, N.J.: Prentice-Hall, 1981, pp. 211–239.

[2] Frederick Jahnel, "Cheating with Charts," Chapter VI in Rudolph Modley and Dyno Lowenstein, *Pictographs and Graphs: How to Make and Use Them,* New York: Harper and Brothers, 1952, pp. 74–75.

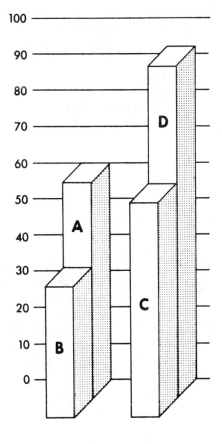

Figure 8-1. The perception of distance can frequently be illusory in three-dimensional charts. For example, Column A may be perceived as taller than Column C. Actually A and C are identical in height. (From Rudolf Modley and Dyno Lowenstein, Pictographs and Graphs, How to Make and Use Them, *New York: Harper and Brothers, 1952, p. 74. Copyright 1952 by Harper and Row, Publishers, Inc. By permission of the publisher.)*

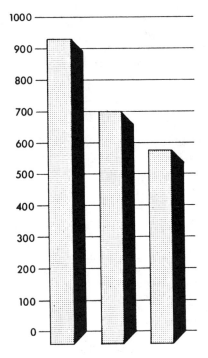

Figure 8-2.   *Because the three columns in this chart have been drawn incorrectly, they give a wrong impression of their height. Apparently, an attempt was made to cast the three columns in perspective or simulated perspective projection. In oblique projection, the slope at the top of the columns would be upward, not downward as shown in the chart. Accordingly, the height of the columns should be measured with reference to their rear dimension as indicated by the scale.* (From Rudolf Modley and Dyno Lowenstein, Pictographs and Graphs, How to Make and Use Them, *New York: Harper and Brothers, 1952, p. 74. Copyright 1952 by Harper and Row, Publishers, Inc. By permission of the publisher.*)

ARE PROJECTION TECHNIQUES TOO TROUBLESOME AND UNRELIABLE TO USE IN STATISTICAL GRAPHICS?

Because of the pitfalls, complexities, and misapplications of three-dimensional charts, there are those who would abandon this graphic form altogether. Such a view is shortsighted and self-defeating since three-dimensional charts, when properly designed and executed, do have a significant role to play in statistical graphics. A rational approach to an issue of this kind involves certain pragmatic considerations that are applicable to all types of charts. In the process of selecting a statistical chart, given certain conditions, requirements, and objectives, one of the most important decisions is which type of chart would be most helpful in doing the best job. The choice may be among an arithmetic line chart, a semilogarithmic chart, a column chart, a surface chart, or some other type of chart. Furthermore, consideration may be given to the advantages and disadvantages of casting the chart that has been selected into three-dimensional form. In a particular instance, for example, if emphasis were placed on "eye appeal" or "novelty," a three-dimensional chart might be given serious consideration since "eye appeal" and novelty are two positive features of three-dimensional charts. How-

ever, a three-dimensional chart should seldom, if ever, be selected for these reasons alone. As pointed out previously, the overall effectiveness of a chart as a vehicle of visual communication depends on many factors and characteristics that include not only appeal, but also clarity, simplicity, accuracy, forcefulness, and interpretability.

The uncertain reputation which three-dimensional charts have acquired seems to be a by-product of the numerous horrendous examples that have found their way into print. However, in this respect, three-dimensional charts are not unique. One is not obliged to search very far in order to find an abundance of egregiously designed and executed "flat" conventional charts. The numerous instances of inferior arithmetic line charts, bar charts, and column charts that have been selected for illustrative purposes in this book attest to this fact. Most, if not all, charts, including inferior three-dimensional charts, are basically attributable to the ignorance and incompetence of the chartmaker, not to the type or character of the chart per se.

Admittedly, in this connection, it can be said that considerably more skill and experience are required in the preparation of acceptable three-dimensional charts than in the preparation of traditional charts. As indicated in the introductory portion of this chapter, genuine expertise of this kind requires a compre-

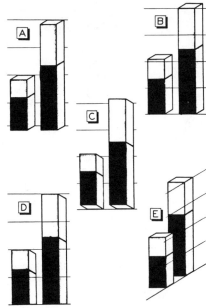

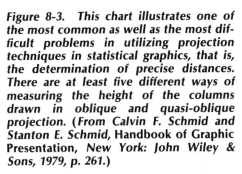

Figure 8-3. This chart illustrates one of the most common as well as the most difficult problems in utilizing projection techniques in statistical graphics, that is, the determination of precise distances. There are at least five different ways of measuring the height of the columns drawn in oblique and quasi-oblique projection. (From Calvin F. Schmid and Stanton E. Schmid, Handbook of Graphic Presentation, *New York: John Wiley & Sons, 1979, p. 261.)*

hensive knowledge of chart design in addition to a thorough understanding of the theory and practice of projection technique.

ILLUSTRATIONS OF PITFALLS, ERRORS, AND DEFICIENCIES IN THREE-DIMENSIONAL CHARTS

A Few Problems Pertaining to the Measurement and Perception of Distance

Frederick Jahnel demonstrates in Figure 8-1 how perceptions of distance can be obscured and distorted by the placement of columns and other symbols as part of a three-dimensional chart.[2] In a comparison of the heights of Columns A and C, one may properly ask, Which is taller? The manner in which the columns have been drawn with their depth and shadow, as well as their location, particularly with reference to the horizontal scale lines, makes Column A seem taller. Actually, Columns A and C are identical in height.

Jahnel presents in Figure 8-2 another illustration in an attempt to show that the "receding depth (in the columns) gives the impression that the amounts are 30 units higher than they actually are. The first block (column), under 900 units high, is made to appear like 930."[3] The "receding depth" in the columns as well as other discrepancies are the result of using the wrong type of projection combined with errors in de-

sign. Apparently, in constructing these columns, an attempt was made to simulate perspective projection. Perspective or quasi-perspective projection is not appropriate for a chart of this kind. If oblique projection had been chosen, the slope at the top of the columns would have been the same as the slope at the bottom, and the discrepancies between the front of the columns and the scale would not have occurred.

Obviously from both the designer's and user's standpoint, the determination of dimensions in three-dimensional charts is extremely important. Generally, whenever practicable as well as compatible with good design principles, scales should be used. Figure 8-3 illustrates five different ways of measuring height in a column chart that has been drawn in oblique projection. Judged in terms of precision and clarity, all of the five techniques in Figure 8-3 are not equally acceptable. The author considers B the most acceptable of the five techniques.

The Hazards and Limitations of Three-Dimensional Pie Charts

Figure 8-4 exemplifies certain pitfalls in designing a three-dimensional pie chart. The type of projection

---

[3] Frederick Jahnel, "Cheating with Charts," Chapter VI in Rudolph Modley and Dyno Lowenstein, *Pictographs and Graphs: How to Make and Use Them,* New York: Harper and Brothers, 1952, p. 74.

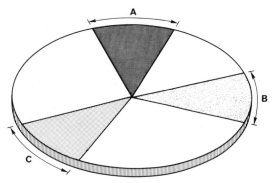

Figure 8-4. A pie chart in isometric projection. Each of the three shaded sectors represents an equal number of degrees on the arc of the ellipse. Note the distortions and variations in size.

used as well as the number and location of the sectors in the "pie," influence the shape and size of the sectors. The basic geometric form of Figure 8-4 is an isometric ellipse with three shaded sectors representing equal degree values. Each of the three shaded divisions is 40° on the arc of the ellipse, but unlike a circle the size and shape of the sectors are distorted and distinctly different from one another. In addition to the three shaded sectors, there are three other sectors of varying size.

Obviously, the more a projected form deviates from circularity, the greater the distortions of the various sectors. Because of the pronounced distortions reflected by Figure 8-4, it is apparent that an isometric projection is unsuitable for a pie chart with so many sectors.

Figure 8-5 is also an isometric ellipse, but with only two major divisions. Judged in terms of comparative area and distance on the arc of the ellipse, the distortions and discrepancies shown by the two sectors may be viewed as minor ones, and the chart is generally acceptable.

Figure 8-6 has been included as another illustration of the pitfalls of designing three-dimensional pie charts with numerous sectors. This chart has 10 sectors, which reflect many discrepancies and distor-

tions.[4] For example, the sectors representing advertising (3 percent), depreciation (3 percent), and rent (3 percent) are all unequal in size, and the sector representing business taxes (4 percent) is smaller than the one representing rent (3 percent). Again, the arc of the sector for packaging (13 percent) is noticeably larger than the combined arcs of corporate profits (6 percent), interest, repairs, and so on (4 percent) and other (5 percent) which total 15 percent).

Figure 8-7 may be thought of as at least one solution to the discrepancies and distortions of three-dimensional pie charts such as the one shown in Figure 8-6. The figures in both charts are comparable, and the categories are identical. The basic geometric form of Figure 8-7 is a true circle with special shading to simulate three-dimensional projection. In measuring degrees on a circle, identical values are represented by identical distances, with no adjustments that might create discrepancies or distortions.

Perspective Projection Can Be Dangerous

As indicated in the introductory portion of this chapter, among the basic forms of projection—axonometric, oblique and perspective—the least applicable and

AVERAGE MONTHLY HOUSEHOLD INCOME OF FOOD STAMP RECIPIENTS

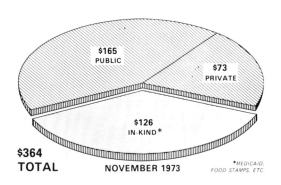

Figure 8-5. A pie chart in the form of an isometric ellipse that has been divided into two major sectors. The distortions in a projection of this kind are relatively minor in comparison to those of similar charts with several sectors. See Figures 8-4 and 8-6 for comparison. [From United States Department of Agriculture, Handbook of Agricultural Charts, 1975, Agricultural Handbook No. 491, 1975, p. 73.]

most problematic in statistical graphics is perspective projection. This kind of projection is particularly troublesome in all time-series charts such as rectilinear coordinate line charts, column charts, and surface charts. Although Figure 8-8 is only a simulation of true perspective projection, it emphasizes the characteristic weaknesses of this type of projection. The tapering grid, unequal spacing of time intervals, and other distortions simply make it impossible to inter-

dards. To begin with, the spacing of the intervals on the horizontal axis is incorrect. The spacing of the first two intervals, each representing two-year periods, is satisfactory, but the third interval with the same spacing includes over three years, and the last interval, showing only about seven months, covers a space almost 30 percent larger than the preceding one. An examination of the vertical measurements of the several surfaces reveals many more strange in-

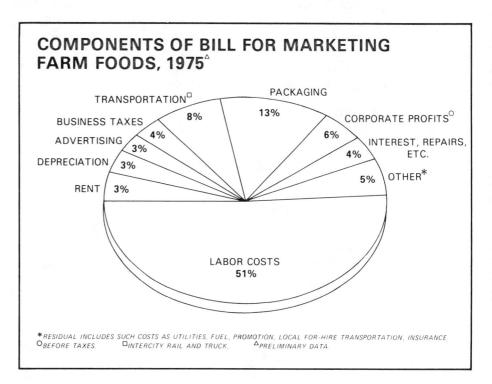

Figure 8-6. Although this pie chart manifests greater circularity than the isometric projections in Figures 8-4 and 8-5, its 10 sectors reflect clearly observable discrepancies and distortions. Note, for example, the sectors labeled "advertising," "depreciation," and "rent," each with 3 percent. See text for further comments. [From United States Department of Agriculture, Handbook of Agricultural Charts, 1976, Agricultural Handbook No. 504, 1976, p. 33.]

pret accurately and meaningfully a chart of this kind. Wherever precise measurement is required, perspective projection should be sedulously avoided. This caveat embraces virtually all types of statistical charts. However, the present writer has found at least one appropriate as well as significant use of perspective projection in designing certain kinds of base maps. This application will be discussed later in this chapter.

Attractive Charts Can Be Misleading

Although the basic design features of Figure 8-9 may seem attractive and authentic, most of the measurements and the manner in which the surfaces have been laid out, do extreme violence to graphic stan-

consistencies and errors. In fact, it seems that each of the scales for the five surfaces is entirely independent of the others, and comparisons in terms of dollars and cents are simply impossible—an extraordinarily bizarre method of statistical charting. For example, the surface for "Alka-Seltzer" begins with $.52, "running shoes," $19.90, and "10-gallons gas," $3.60. This indicates that $19.90 is over 38 times larger than the price of "Alka-Seltzer," but the height of surface representing "running shoes" is about 1.4 times as high as the surface of the 1972 price of "Alka-Seltzer." Similarly, the price of "10-gallons gas" in 1972 was $3.60, which was almost seven times as much as the price of "Alka-Seltzer," but the respective heights of

---

[4] Incidentally, any type of pie chart with as many as 10 sectors is usually unsatisfactory.

the surfaces are reversed. That is, the surface for "10-gallons gas" is about six-tenths as high as the surface for "Alka-Seltzer" in 1972. A further examination of this chart will reveal additional errors and discrepancies.[5]

Eccentricities and Complexities in the Design of Statistical Charts Can Create Serious Deficiencies

The design of Figure 8-10 is a combination of two kinds of bar charts: the area bar chart and the subdivided bar chart. The face of the chart as measured by the vertical scale, indicates the percentage of lost

workday cases due to injuries and illnesses (injuries and illnesses resulting in 15 or more days away from work) in accordance with the size (number of employees) of the establishment. The side of the chart, as measured by the horizontal scale, is composed of a series of subdivided area bar charts that show the proportion of cases involving 15 or more days away from work, and the proportion involving less than 15 days of work classified by size (number of employees) of the establishment. For example, establishments with 20–40 employees reported that 26.2 percent of the cases were away from work 15 or more days because of injuries and illnesses and that 73.8 percent of the cases involved less than 15 days away from work.

Although technically correct as an example of

WHAT MAKES UP THE FARM-FOOD MARKETING BILL

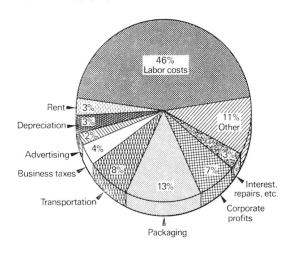

Figure 8-7. A three-dimensional pie chart in which circularity has been maintained. As far as the several sectors are concerned, there are no discrepancies in shape or size. It will be observed that this chart presents statistics similar to those in Figure 8-6, but for 1977 rather than for 1975. [From United States Department of Agriculture, Hand-book of Agricultural Charts, 1978, Agricultural Handbook No. 551, 1978, p. 52.]

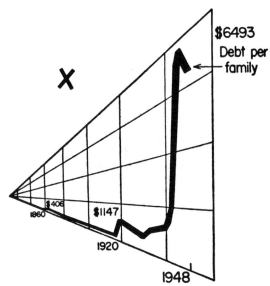

Figure 8-8. Although crudely drawn and impressionistic, this chart illustrates some of the dangers of perspective projection. The tapering grid, unequal spacing of time intervals, and other distortions characterize all rectilinear coordinate time charts drawn in perspective. (From W. Allen Wallis and Harry V. Roberts, Statistics, A New Approach, Clencoe, Illinois: The Free Press, 1956, p. 86. Copyright 1956 by The Free Press.)

oblique projection, this chart fails to meet the requirements of an effective medium of visual communication. First, because of its complexity and eccentricity, it is not readily interpretable. In fact, judging from the experience and knowledge of most readers, it is doubtful if they could derive much meaning from this chart. Second, one particularly serious limitation of this chart is the vertical scale, which is supposed to relate to the data on the face of the bars. The resultant visual impression between the scale and the face of the bars is one of distortion, distance, and unrelatedness. Such characteristics in a statistical chart as novelty, ingenuity, originality, and drafting expertise are irrelevant and inconsequential if the chart is not clear and readily understood.

Novelty Alone Is Not a Sound Basis for Chart Design

Figure 8-11 is a strange and seemingly irrelevant application of three-dimensional techniques for portraying several series of monthly accident and illness rates by industry division on a semilogarithmic scale.

---

[5] Many of the "eye-catching" and "flashy" charts published in newspapers are seriously flawed. For an informative and critical discussion of this subject, see: Howard Wainer, "Making Newspaper Graphs Fit to Print," in Paul A. Kolers, Merald E. Wrolstad, and Herman Bouma (eds.), *Processing of Visible Language 2,* New York: Plenum, pp. 125-142.

Figure 8-9.   A surface chart in trimetric projection. This chart is seriously marred by the incorrect spacing of the time intervals and the grossly inconsistent and uncertain measurements of the several surfaces. (From the Seattle Times, Thursday, November 15, 1979, p. D2.)

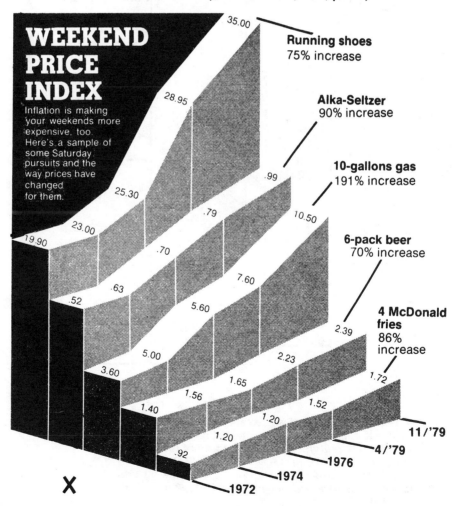

Percent distribution of lost workday INJURIES AND ILLNESSES, and percent of lost workday INJURIES AND ILLNESSES involving 15 or more days away from work, by employment-size group, private sector, United States, 1975

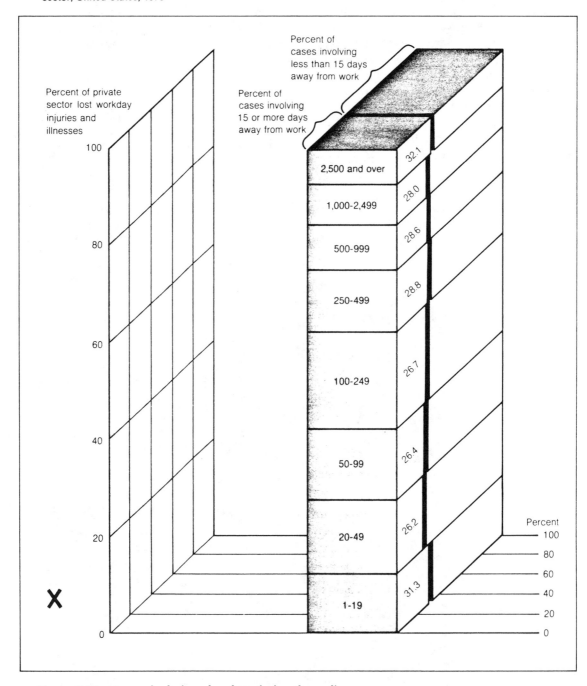

Figure 8-10. A poorly designed and confusing three-dimensional chart. Perhaps the most difficult feature of this chart is the interpretation of the face of the column as it relates to the vertical scale. [From U.S. Department of Labor, Bureau of Labor Statistics, Chartbook on Occupational Injuries and Illnesses in 1975, Report 501, 1977, p. 4.]

Injury and Illness Incidence Rates by Month, by Industry Division, United States, 1974

Incidence rates calculated on a monthly basis show modest seasonal variation in all industry divisions, with the highest rates in the summer months and the lowest rates at the end of the year.

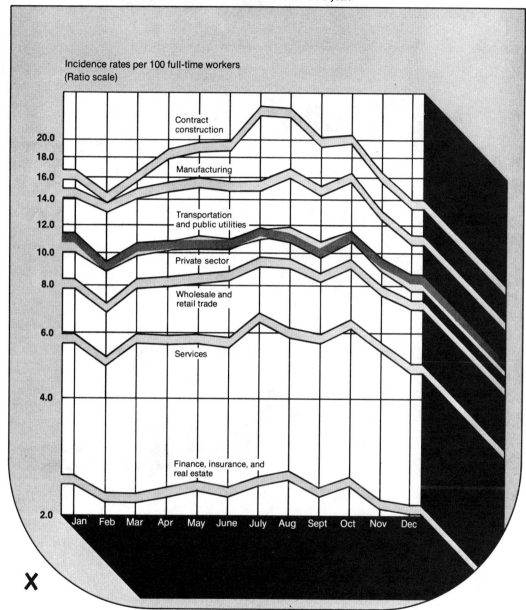

Incidence rates per 100 full-time workers
(Ratio scale)

Contract construction

Manufacturing

Transportation and public utilities

Private sector

Wholesale and retail trade

Services

Finance, insurance, and real estate

X

Figure 8-11. A three-dimensional chart of this kind is difficult to justify. It attempts to portray seven series of monthly data on a semilogarithmic grid. The third dimension shading along with the extension of the curves merely cause clutter and possible confusion. [*From U.S. Department of Labor, Bureau of Labor Statistics,* Chartbook on Occupational Injuries and Illnesses, 1974, *Report 460, 1976, p. 9.*]

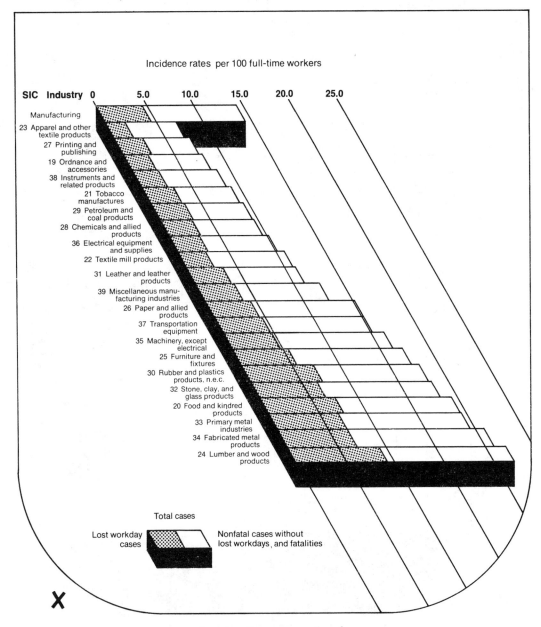

Injury and Illness Incidence Rates by Type of Manufacturing Activity, United States, 1974

Incidence rates in manufacturing industries ranged from 7.1 in apparel and other textile products to 22.2 in lumber and wood products.

Incidence rates per 100 full-time workers

SIC Industry

Manufacturing
23 Apparel and other textile products
27 Printing and publishing
19 Ordnance and accessories
38 Instruments and related products
21 Tobacco manufactures
29 Petroleum and coal products
28 Chemicals and allied products
36 Electrical equipment and supplies
22 Textile mill products
31 Leather and leather products
39 Miscellaneous manufacturing industries
26 Paper and allied products
37 Transportation equipment
35 Machinery, except electrical
25 Furniture and fixtures
30 Rubber and plastics products, n.e.c.
32 Stone, clay, and glass products
20 Food and kindred products
33 Primary metal industries
34 Fabricated metal products
24 Lumber and wood products

Total cases

Lost workday cases   Nonfatal cases without lost workdays, and fatalities

X

Figure 8-12. *Another example of a three-dimensional chart which does not exhibit any special qualities that make it superior to the conventional composite bar chart as a visual medium of communication.* [*From U.S. Department of Labor, Bureau of Labor Statistics,* Chartbook on Occupational Injuries and Illnesses, 1974, *Report 460, 1976, p. 6.*]

The black shading, the extra spacing on the grid, and the unnecessarily broad width and extended length of the seven curves merely add to the clutter and visual confusion of the chart. This figure shows how futile resorting to pseudo-three-dimensional techniques merely for the purpose of achieving novelty and attracting attention can be.

An Additional Illustration of the Irrelevancy and Inappropriateness of a Three-Dimensional Chart

Figure 8-12 is another example of a misapplication of three-dimensional projection. The basic graphic form is a relatively simple subdivided or composite bar chart. It shows accident and illness rates for a number of different types of manufacturing industries. A careful examination of this chart could very well elicit the following question: Why was this chart designed in three-dimensional form rather than conforming to the customary conventional type? One might argue that the three-dimensional version has more "eye appeal" and because of its "novelty" attracts more attention. However, although Figure 8-12 is definitely more novel than a conventional "flat" bar chart, novelty merely for the sake of novelty or for "eye appeal" alone does not represent a justifiable rationale for selecting a three-dimensional chart or any other type of chart. To select an appropriate chart design, one must

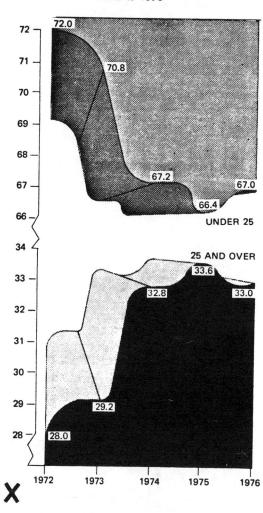

Age structure of college enrollment: 1972 to 1976

Figure 8-13. A strange attempt to cast two simple series of data into a three-dimensional chart. The results are confusing as well as unattractive. (From W. Vance Grant and C. George Lind, Digest of Education Statistics: 1979, *United States Department of Health, Education and Welfare, National Center for Education Statistics, 1979, p. 97.)*

SOURCE: U.S. Department of Commerce, Bureau of the Census, *Current Population Reports,* Series P-20, numbers 260, 272, 286, 303, and 307.

answer the following significant and fundamental questions: Does this type of chart do a much better job than the conventional chart? Is it clearer? More readily interpretable? More attractive? More accurate and reliable? More effective visually? These are the types of questions that must be addressed before a sound decision can be made.

The Type of Chart Selected as Well as the Design Should Be Appropriate: Another Example Where Both Have Failed

Figure 8-13, which purports to depict trends in the age structure of college enrollment from 1972 to 1976, is another example of a poorly designed three-dimen-

sional surface chart. In terms of design standards, it is grossly defective. The broken and idiosyncratic scale, the strangely projected scale lines, the bizarre rounding of the curves, and the distortions representing the third dimension constitute specific design errors and deficiencies. As a statistical chart, it is a failure. It is not readily interpretable, accurate, clear, simple, forceful, or attractive.

Another Illustration of an Inappropriate and Poorly Designed Three-Dimensional Surface Chart

Figure 8-14 is a mediocre attempt to portray trends in the educational status of the population by means of

Figure 8-14.   A poorly designed surface or stratum chart in three dimensions. The difficulty of making vertical measurements, the virtual irrelevance of the ordinal scale, and the makeshift crossover of the two smaller strata are three defects. (From W. Vance Grant and C. George Lind, Digest of Education Statistics: 1979, United States Department of Health, Education and Welfare, National Center for Education Statistics, 1979, p. 17.)

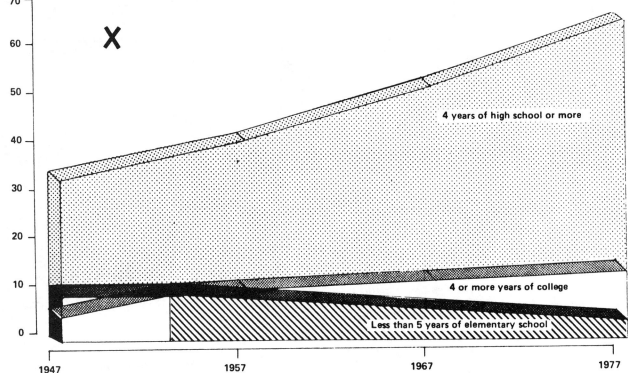

·Level of school completed by persons 25 years old and over: United States, 1947 to 1977

NOTE.—Data are based upon sample surveys which include the civilian noninstitutional population and members of the Armed Forces living off post or with their families on post.

SOURCE: U.S. Department of Commerce, Bureau of the Census, *Current Population Reports,* Series P-20, Nos. 15, 77, 169, and 314.

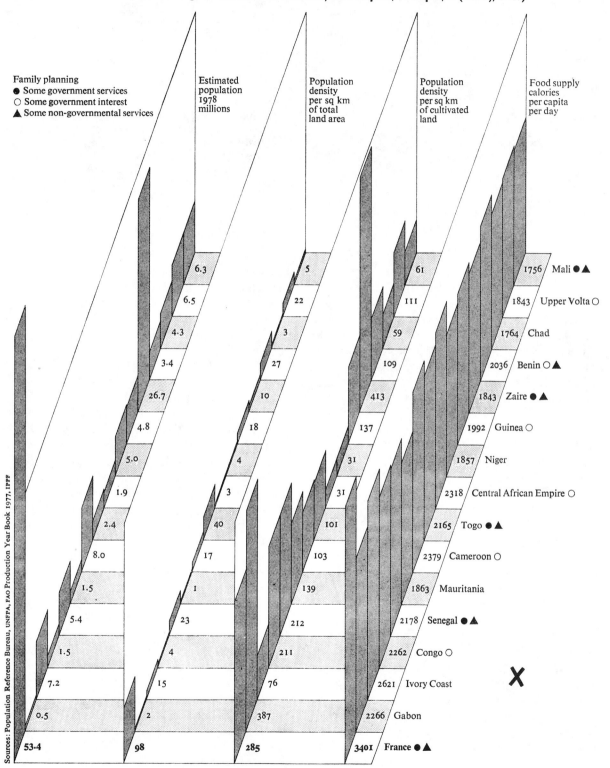

Figure 8-15. Because of its detail, complexity, and generally poor design, this chart is not an effective device of visual communication. See text for further comments. (*From "Lingua Franca for a Diversity of Peoples," People, 6 (1979), 3–6.*)

Family planning
● Some government services
○ Some government interest
▲ Some non-governmental services

Estimated population 1978 millions

Population density per sq km of total land area

Population density per sq km of cultivated land

Food supply calories per capita per day

Sources: Population Reference Bureau, UNFPA, FAO Production Year Book 1977, IPPF

Family planning	Est. pop. 1978 (millions)	Pop. density total land	Pop. density cultivated land	Food supply cal/capita/day	Country
	6.3	5	61	1756	Mali ● ▲
	6.5	22	111	1843	Upper Volta ○
	4.3	3	59	1764	Chad
	3.4	27	109	2036	Benin ○ ▲
	26.7	10	413	1843	Zaire ● ▲
	4.8	18	137	1992	Guinea ○
	5.0	4	31	1857	Niger
	1.9	3	31	2318	Central African Empire ○
	2.4	40	101	2165	Togo ● ▲
	8.0	17	103	2379	Cameroon ○
	1.5	1	139	1863	Mauritania
	5.4	23	212	2178	Senegal ● ▲
	1.5	4	211	2262	Congo ○
	7.2	15	76	2621	Ivory Coast
	0.5	2	387	2266	Gabon
53.4	98	285	3401		France ● ▲

X

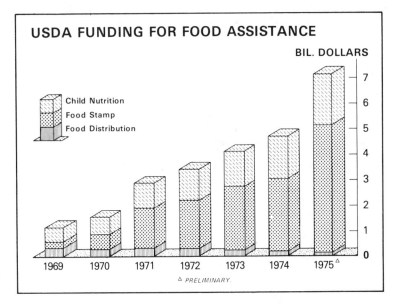

USDA FUNDING FOR FOOD ASSISTANCE

BIL. DOLLARS

Child Nutrition
Food Stamp
Food Distribution

1969   1970   1971   1972   1973   1974   1975△

△ PRELIMINARY.

Figure 8-16. A subdivided column chart in oblique projection. This chart has a scale that is gauged to the face of the respective columns. [From United States Department of Agriculture, Handbook of Agricultural Charts, 1975, Agricultural Handbook No. 491, 1975, p. 72.]

a surface chart in three dimensions. The fact that the data do not lend themselves very well to a three-dimensional design of this kind apparently created a problem for the chartmaker. However, the projection of the three strata is not correct and the improvisation of the crossover in the data is poorly executed. In light of these facts, it would have been much more appropriate if the chartmaker had designed a conventional line chart or column chart.

A Detailed, Highly Complex and Elaborate Three-Dimensional Chart: It Fails to Meet the Standards of an Effective Medium of Visual Communication

In comparison to other charts in this chapter, Figure 8-15 is one of the more complicated and detailed. In some respects it is a combination of a statistical table and a statistical chart. Comparative data for 15 African nations and for France are shown in the chart. The data include (1) size of population in 1978; (2) population density based on total land area; (3) population density based on cultivated land area; and (4) a food supply index based on calories per capita per day. Additional information pertaining to type of family planning services is indicated by symbols. The chart embodies an extraordinary amount of information, but its graphic qualities manifest serious limitations. The major graphic features of this chart are four series of column charts that attempt to portray comparisons of the four basic indexes described ear-

lier. However, there are no scales for the columns and the columns are distorted. The figures at the base of each of the 64 columns indicate the values represented by the columns.

ILLUSTRATIONS OF GENERALLY ACCEPTABLE CHARTS DRAWN IN THREE DIMENSIONS

Figure 8-16 is a well-designed subdivided or component column chart drawn in oblique projection. The base line of the scale is coincidental with the face of the columns. However, the total height measurements, either along the face or along the back of the columns, are identical. The three components are clearly differentiated by an appropriate hatching scheme.

In order to avoid the problem of scale rulings, at least for simple column charts drawn in oblique projection, it is a fairly common practice to omit scale lines altogether. A total figure is placed at the top of the column. Also, it will be found that flat, conventional column charts are sometimes designed in this manner, although it is easier and perhaps better practice to include a scale.

Figure 8-17 is an example of a column chart without a scale drawn in oblique projection. It portrays student enrollment as measured by two different indexes for certain institutions and groups of institutions of higher learning in the state of Washington. The face of each column is the basis for determining height. In designing the chart, the importance of the

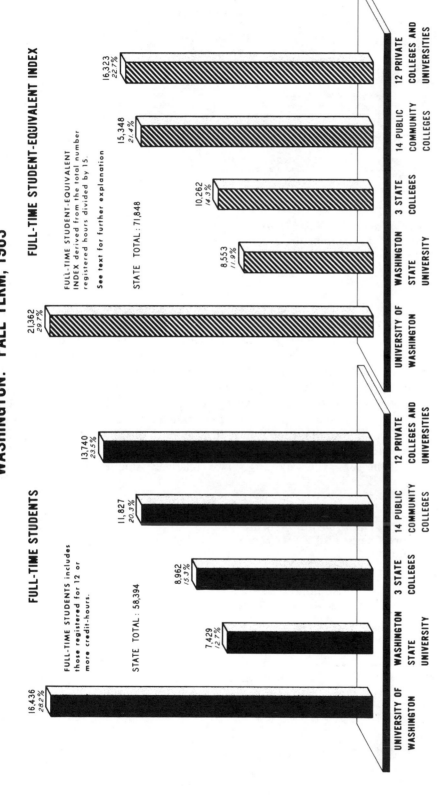

FULL-TIME STUDENTS AND FULL-TIME STUDENT-EQUIVALENT INDEX
WASHINGTON: FALL TERM, 1963

FULL-TIME STUDENTS

FULL-TIME STUDENTS includes
those registered for 12 or
more credit-hours.

STATE TOTAL: 58,394

UNIVERSITY OF WASHINGTON — 16,436 28.2%
WASHINGTON STATE UNIVERSITY — 7,429 12.7%
3 STATE COLLEGES — 8,962 15.3%
14 PUBLIC COMMUNITY COLLEGES — 11,827 20.3%
12 PRIVATE COLLEGES AND UNIVERSITIES — 13,740 23.5%

FULL-TIME STUDENT-EQUIVALENT INDEX

FULL-TIME STUDENT-EQUIVALENT
INDEX derived from the total number
registered hours divided by 15.

See text for further explanation

STATE TOTAL: 71,848

UNIVERSITY OF WASHINGTON — 21,362 29.7%
WASHINGTON STATE UNIVERSITY — 8,553 11.9%
3 STATE COLLEGES — 10,262 14.3%
14 PUBLIC COMMUNITY COLLEGES — 15,348 21.4%
12 PRIVATE COLLEGES AND UNIVERSITIES — 16,323 22.7%

Figure 8-17. A column chart without a scale drawn in oblique projection. The face of the respective columns has been used as the determinant of height. Note the figures at the top of each of the columns. [From Calvin F. Schmid, Vincent A. Miller, and William S. Packard, Enrollment Statistics, Colleges and Universities, State of Washington, Fall Term, 1963, Seattle: Washington State Census Board, 1964, p. 9.]

USDA COST OF THE FOOD STAMP PROGRAM

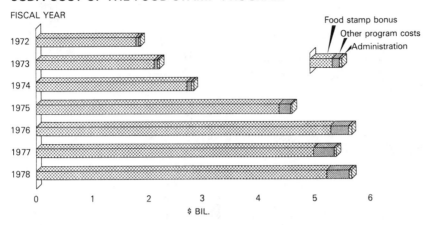

Figure 8-18. A composite or subdivided bar chart in oblique projection. See text for comments. [From United States Department of Agriculture, Handbook of Agricultural Charts, 1978, Agricultural Handbook No. 551, 1978, p. 63.]

face of the columns, in contrast to the other characteristics of the columns, has been emphasized. The figures at the top of each column indicate both the numerical and percentage distribution of student enrollment for each of the respective categories. In subdivided or component column charts, this design may prove unsuitable because of clutter and the difficulty of estimating values of the various components. Figure 8-18 is a subdivided or component bar chart in oblique projection. The main referent line for the scale coincides with the face of the bars. The scale consists merely of a series of figures without scale points or scale lines. It will be noted that the front and rear dimensions of each of the bars are identical. Since the data portrayed in this chart consist of a time series, it might have been preferable to have selected a column chart rather than a bar chart. Of course, whether the graphic form selected in this instance is a bar or column chart, the design techniques, particularly in relation to problems of projection and measurement, are very similar.

Two relatively simple bilateral bar charts in

Figure 8-19. Two bilateral bar charts in oblique projection. [From United States Department of Agriculture, Handbook of Agricultural Charts, 1978, Agricultural Handbook No. 551, 1978, p. 118.]

Changes in Frozen Vegetable Consumption per Capita Between 1970–72 and 1976–78

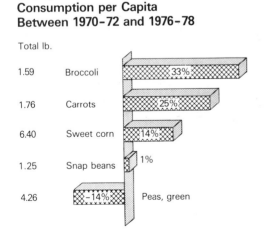

Changes in Canned Vegetable Consumption per Capita Between 1970–72 and 1976–78

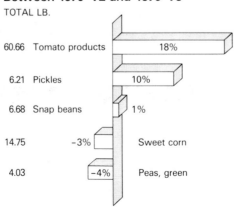

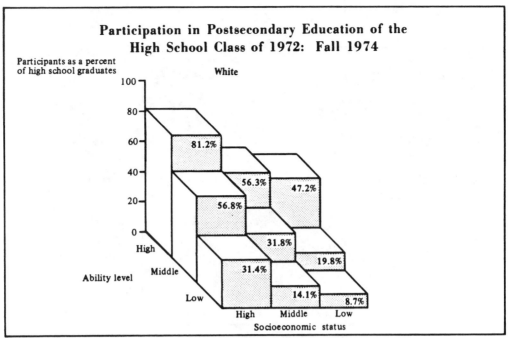

Participation in Postsecondary Education of the High School Class of 1972: Fall 1974

Participants as a percent of high school graduates

White

Ability level

Socioeconomic status

Source of Data: National Center for Education Statistics, National Longitudinal Study

Figure 8-20. A three-dimensional block diagram that shows how participation in postsecondary education is related to ability level and to socioeconomic status. The highest participation, as measured as a percentage of high school graduates, is indicated by the group that ranks highest in both ability and socioeconomic status (81.2 percent). By contrast, the group that ranks lowest in ability and in socioeconomic status shows the smallest proportion (8.7 percent) in postsecondary education. Although some of the blocks are partially concealed and are difficult to measure with exactitude, the scale and percentage labels make a chart of this kind acceptable with respect to both graphic qualities and reliability. (From Mary A. Golladay, The Condition of Education, United States Department of Health, Education, and Welfare, National Center for Education Statistics, 1977, p. 86.)

Figure 8-21. A subdivided or component column chart represented by a series of three-dimensional symbols in cylindrical forms. The chart portrays agricultural and nonagricultural water use from 1960 to 1975. The basic symbol is in the form of a cylindrical reservoir or possibly a dug well. Although the chart has a scale, vertical distances are difficult to measure with any degree of exactitude, because of the curvatures both at the bottom and at the top of each cylinder. (From United States Department of Agriculture, 1978 Handbook of Agricultural Charts, Agricultural Handbook No. 551, Washington, D.C.: Government Printing Office, 1978, p. 29.)

WATER USE

MIL. ACRE FEET

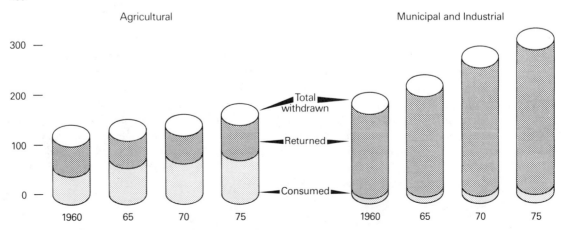

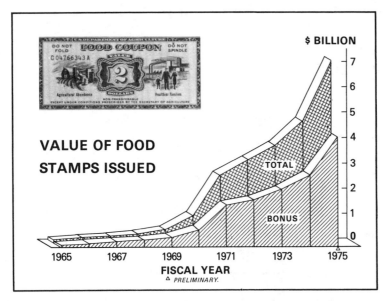

Figure 8-22. A chart with two surfaces and one scale drawn in oblique projection. The zero base line coincides with the innermost side of the surface labeled "total." Because of the thickness of the surfaces and their angular recession, exact vertical measurements from the base line are not possible. [From United States Department of Agriculture, Handbook of Agricultural Charts, 1975, Agricultural Handbook No. 491, 1975, p. 75.]

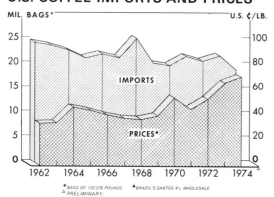

Figure 8-23. A surface chart with two scales drawn in oblique projection. The chart pertains to coffee imports with volume measured in bags and price measured in amount per pound. The scales are gauged from the innermost dimension of the surface that represents volume of imports. The downward slope of the two surfaces from the zero base line creates inevitable discrepancies in vertical measurements. [From United States Department of Agriculture, Handbook of Agricultural Charts, 1975, Agricultural Handbook No. 491, 1975, p. 157.]

oblique projection are shown in Figure 8-19. Neither of these charts has scales. The faces of each bar, both positive and negative, are commensurate with the percentages indicated in the body of each chart. The outer borders of the dividers between positive and negative values represent the main referent lines.

The inner dimensions of the respective positive bars are measured from the inner dimensions of the dividers between positive and negative values and are the same length as the face of the respective bars. In the case of the three negative bars, the inner dimensions are determined in the same manner, but their full inside lengths are partially concealed. This fact could be judged a deficiency of this type of chart, but certainly it is not a defect of major proportions since, it will be recalled, the faces of both positive and negative bars are comparable and can be determined fairly reliably.

Age at first attack in RHEUMATIC FEVER

ALL AGES

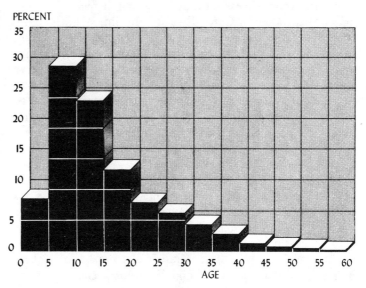

2,535 cases, Philadelphia Hospitals, 1930-1934

(after Hedley)

**CHILDHOOD
AND YOUNG
ADULT AGES**

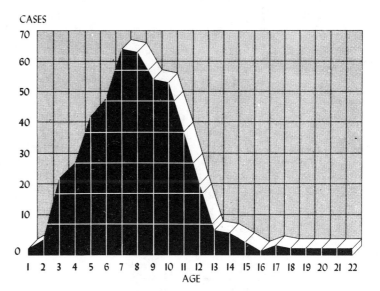

500 cases, New Haven Hospital and Dispensary

(after Leonard)

Figure 8-24. Two simple frequency curves cast in three-dimensional form. The upper chart is a histogram, and the lower chart, a frequency polygon. (From Metropolitan Life Insurance Company, Studies in Rheumatic Fever, New York, 1944, p. 9.)

The portrayal of data in Figure 8-20 extends beyond the simple column chart, which characteristically shows how a specified value—number, percentage, rate or other measure—is related to time or to some qualitative category. In Figure 8-20, a specified value—the percentage of high school graduates who go on for postsecondary education—is related to two major variables, ability level and socioeconomic status, each of which in turn is divided into a threefold classification: high, middle, and low. It is clear from this chart that high socioeconomic status as well as high ability level are positively correlated with participation in postsecondary education.

As in most of the charts in this chapter, the symbol in Figure 8-21 is three-dimensional, but the basis of comparison is one-dimensional. In a pictorial sense, the symbol typifies either a cylindrical reservoir or possibly a dug well containing a variable amount of water over time. Graphically, the basic design is a subdivided or component column chart.

The data indicate agricultural and nonagricultural water use recorded for four quinquennial years from 1960 to 1975. Also, the total amount of water withdrawn for agricultural and nonagricultural uses is dichotomized by a shading scheme into the amount consumed and the amount returned. It will be observed that although nonagricultural users (municipal and industrial users) actually withdraw more water than agricultural users, they return most of it to streams. In contrast, irrigators return only about half the water they withdraw. In evaluating Figure 8-21 on the basis of statistical graphics standards, it is apparent that measurements of distance are cumbersome and uncertain in spite of the fact that there is a vertical scale. The curvatures at both the top and bottom of each cylinder definitely complicate vertical measurements.

Figures 8-22 and 8-23 are carefully prepared three-dimensional surface, stratum, or band charts. It will be noted that Figure 8-23 has two scales. Al-

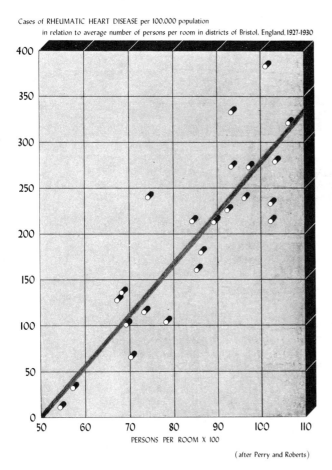

Crowding in Homes Is Closely Related to the Incidence of RHEUMATIC HEART DISEASE

Cases of RHEUMATIC HEART DISEASE per 100,000 population

in relation to average number of persons per room in districts of Bristol, England, 1927-1930

PERSONS PER ROOM X 100

(after Perry and Roberts)

Figure 8-25. A scatter diagram in three-dimensional form. This chart indicates a positive correlation between crowding in homes and rate of rheumatic heart disease. (From Metropolitan Life Insurance Company, Studies in Rheumatic Fever, New York, 1944, p. 13.)

though both of the charts are clear, attractive, and relatively simple, they do exhibit one seemingly unresolvable limitation, that is, discrepant vertical measurements. All vertical measurements in these two charts are made from the zero base line along the inner surface of the two strata or bands. Since the back side of the inside stratum coincides with the zero base line and other parts of the scale, it is only logical that this procedure be followed for this stratum. However, the question that arises is, What line or base should be used as a referent for the outer stratum? There are two alternatives: the zero base line of the vertical scale or some other line on the flat projected plane at the bottom of the chart. For simplicity, and possibily comparability, the designer(s) of these charts chose the first alternative. It should be

pointed out that the projected recession representing the thickness of the two strata offers some compensation for measurement discrepancies. Before selecting three-dimensional charts of this kind, at least the following three questions should be resolved: How large precisely are the measurement discrepancies? Are they within tolerable limits? Do other positive features of this type of chart compensate for relatively small measurement discrepancies?

Figure 8-24 illustrates how simple frequency charts can be cast into three-dimensional form. A histogram is shown at the top and a frequency polygon at the bottom. As in the two preceding charts (Figures 8-22 and 8-23), the third dimension (depth) has created several apparent discrepancies and uncertainties, especially with respect to vertical mea-

Figure 8-26.   Another application of a map drawn in perspective projection. The chart compares the five largest net streams of migration for students and nonstudents according to nine geographic divisions as defined by the United States Bureau of the Census. The dominant position of the Middle Atlantic Division as a source of student migration, and the importance of the Pacific Division as a goal for nonstudent migrants are clearly observable. (From Calvin F. Schmid and Charles S. Gossman, "Characteristics and Patterns of Student and Non-Student Migration," Proceedings of International Union for the Scientific Study of Population, London, 1969, S6.21–S6.214.)

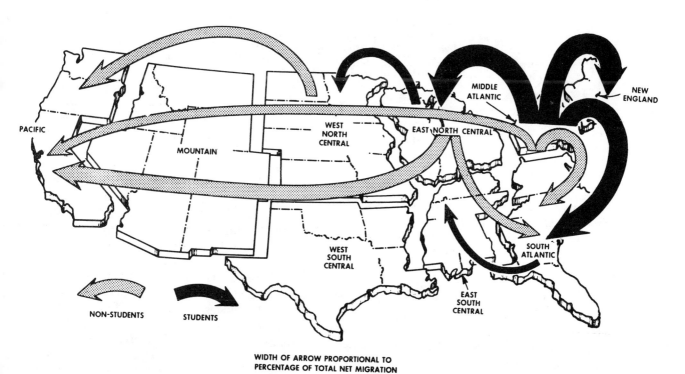

MAJOR STREAMS OF NET MIGRATION

COLLEGE STUDENTS AND NON-STUDENTS : 1955-60

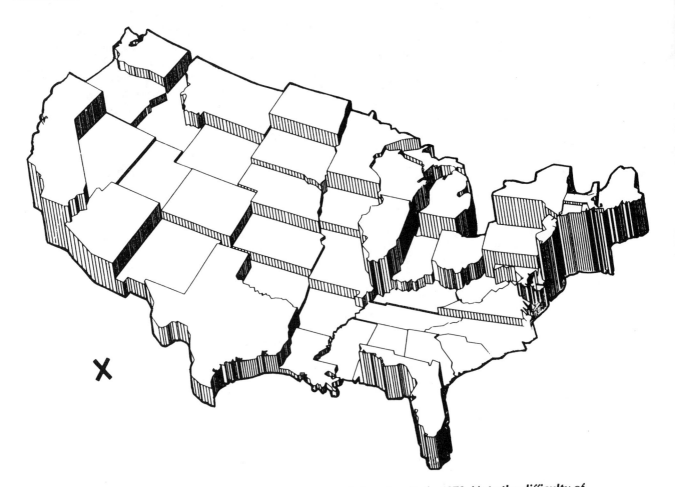

Figure 8-27. A prism map showing population density in 1970. Note the difficulty of measuring, or even estimating, relative density. Also, observe how essential characteristics of a number of states are obscured. (From cover of announcement, Harvard Computer Graphics Week '79, Cambridge, Mass.: Harvard University Laboratory for Computer Graphics and Spatial Analysis, 1979.)

surements. It will be noted that the designer has attempted by projection techniques to integrate the scales both for the back and the face of the histogram and the polygon. This procedure may seem to make the charts more complex by the apparent addition of another grid, while at the same time this procedure gives more precise emphasis to the three-dimensional character of the charts. Because of their limitations and uncertainties, charts of this kind should be selected and designed with extreme care and discrimination.

Figure 8-25 is a scattergram with three-dimensional features. It depicts the positive relationship between crowding in homes and the incidence of rheumatic heart disease. The vertical axis indicates cases of rheumatic heart disease per 100,000 of population, and the horizontal axis, persons per room times 100.

Many forms of three-dimensional maps can be developed into effective statistical charts. One particularly useful type is a base map in perspective projection that is divided into relatively large areas such as states, regions, or administrative areas. The preparation of this type of map can be done in three different ways: (1) manually, by mechanical projection; (2) by electronic computer; or (3) by photography. The base map for Figure 8-26 was prepared photographically. In this process the areas (states, regions, administrative areas, etc.) were cut out with a jigsaw from a map, mounted on at least 1-inch-thick fiberboard or Styrofoam, and then photographed from different heights and angles. The most appropriate photographic reproductions were selected for tracing one or more different base maps. In Figure 8-26, flow lines representing population migration have

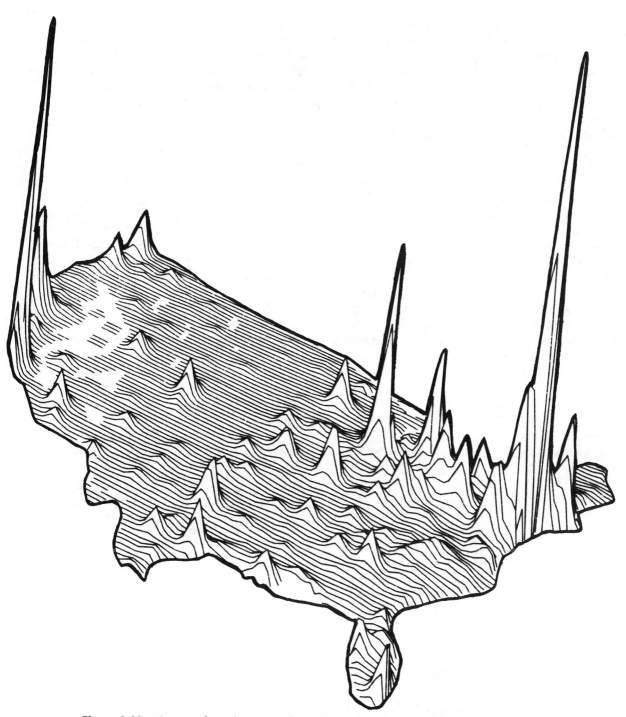

Figure 8-28.   A smooth-surface map depicting population density in 1970. The data for this map are based on counties. See text for additional comments. (From cover of Context, Cambridge, Mass.: Harvard University Laboratory for Computer Graphics and Spatial Analysis, Spring 1979.)

been superimposed on the base map. Similarly, in earlier chapters (Figure 6-8), columns in oblique projection have been superimposed on a base map of the state of Washington, which has 16 administrative areas; in Figure 6-13 spherical symbols and cross-hatching have been superimposed on the two base maps of the United States; in Figure 7-8 spherical symbols have been placed on the single base map of the United States.

At the present time, two of the most popular three-dimensional maps are based on the "statistical surface" concept. They are the "prism" or "block" map and the "smoothed surface" or "stereogram" map.[6] Their popularity may be attributed, at least to a considerable degree, to the prolific numbers prepared from programs and techniques developed by specialists in computer graphics. Up until 1970 or so, most maps of this kind were constructed manually, which was a laborious and time-consuming process. In recent years, the Harvard University Laboratory for Computer Graphics and Spatial Analysis is especially noteworthy for its computer programs and the large number of prism and smoothed surface maps it has produced. Figure 8-27 is an illustration of a prism map. It depicts population density by states for 1970. Although impressive looking, the prism map is inferior to the "flat" choropleth map as a medium of visual communication. It has the same shortcomings as the choropleth map and at least two other limitations as well. First, the height of the prisms are not measurable with any degree of accuracy even though mensurational divisions along with a corresponding scale are sometimes included on the map. Second, in most maps of this kind, depending of course on the number, size, and shape of the areal units, a substantial number of the prisms may be so obscured and distorted by others as to make it impossible to determine their heights or other characteristics.

The smoothed surface map, Figure 8-28, also shows the distribution of population of the United States in 1970. In constructing this map, county data were used. Obviously since there are over 3000 counties in the United States, a prism map based on areal units of such size and number would be most impracticable. As an overall portrayal of population distribution that exhibits a high degree of clarity, specificity, interpretability, and appeal, it is unquestionably superior to the prism map portrayed in Figure 8-27. Also, in comparison to other techniques for showing population distribution of a relatively large area it would be considered an acceptable alternative. Besides the smoothed surface map, the following types of charts are commonly used to portray population distribution: (1) multiple-dot map, (2) isopleth map, (3) choropleth map, (4) graduated three-dimensional symbol map (spheres), and (5) graduated two-dimensional symbol map (circles). Of course, the smoothed surface chart has other applications besides depicting population distribution.

---

[6] For a discussion of the basic theory and application of the statistical surface concept, see: Calvin F. Schmid and Earle H. MacCannell, "Basic Problems, Techniques, and Theory of Isopleth Mapping," *Journal of the American Statistical Association,* **50** (1955), 220–239. Also, see: George F. Jenks, "Generalizations in Statistical Mapping," *Annals, Association of American Geographers,* **53** (1963), 15–26.

BECAUSE OF THEIR POPULAR APPEAL AND their characteristically simple and artistic features, it might be assumed that pictorial charts are relatively free of serious design problems. In actual practice, however, this is not the case. In fact, one frequently used type of pictorial chart where, ostensibly, the size of the symbol is constructed commensurate with the value that is depicted almost invariably does violence to accepted standards and principles of chart design. Of course, this type of pictorial should be carefully avoided. The remaining types of pictorial charts, like many traditional graphic forms, are susceptible to less serious pitfalls and deficiencies.

In general, pictorial charts may be subsumed under four basic types: (1) charts in which the size of the pictorial symbols is made in proportion to the values portrayed; (2) pictorial unit charts in which each symbol represents a definite and uniform value; (3) cartoon and sketch charts in which the basic graphic form, such as a curve or bar, is portrayed as a picture; and (4) charts with pictorial embellishments ranging from a single pictorial filler to an elaborate and detailed pictorial background.[1]

WHEN SHOULD PICTORIAL CHARTS BE USED?

A realistic response to the question, When should pictorial charts be used? depends on a careful evaluation of three basic considerations: (1) the medium in which the chart is to be presented; (2) the education or "knowledge" level of the audience for whom the chart is to be designed; and (3) the nature of the statistics that are to be portrayed.[2]

If the chart is to be used in a popular lecture, in a television program, or in a newspaper or magazine where contact with an audience is relatively ephemeral and possibly superficial, a pictorial chart could be appropriate, especially in the case of newspapers and magazines, where charts would have to compete for attention with other items in the publication.

If an audience or other group of users is of "average" or below average educational background, a well-designed pictorial chart could facilitate the com-

CHAPTER NINE

PICTORIAL CHARTS

Critique and Guidelines

munication of a message as well as serve as a stimulus for eliciting interest and appeal.

This is where the skill and experience of the designer is especially important. His art in designing the symbol largely determines whether or not it will help the reader to understand the chart. In other words, the symbol must be designed simply and yet it should be easily recognized.[3]

The nature or characteristics of the statistical data could be a decisive determinant in the design of an appropriate pictorial chart. In fact, in some instances the data themselves preclude the use of pictorial representation. This might happen if changes in the data are excessively small or excessively large, or markedly anomalous in other respects. When such a problem does occur, it is necessary, of course, either to select a conventional graphic form or dispense entirely with graphic presentation.

AN ABERRANT AND DECEPTIVE TYPE OF PICTORIAL CHART

Of the four different types of pictorial charts described in the initial section of this chapter, the most deficient and misleading is based on the criterion of size. In theory, the size of each symbol is made proportionate to the value being portrayed, but in actual practice this is seldom the case. Almost invariably, the size of the symbols is not measured correctly. Fur-

[1] Calvin F. Schmid and Stanton E. Schmid, *Handbook of Graphic Presentation,* New York: John Wiley & Sons, 1979, p. 220.

[2] Frederick Jahnel, "When to Use Pictorial Symbols," *The American Statistician, 2* (1948), 25.

[3] Frederick Jahnel, "When to Use Pictorial Symbols," *The American Statistician, 2* (1948), 25.

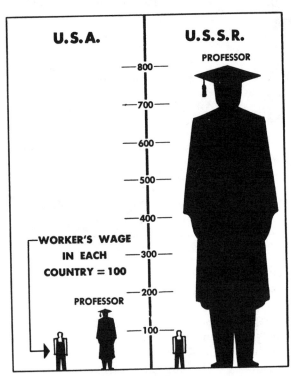

U.S.A.   U.S.S.R.

PROFESSOR

— 800 —
— 700 —
— 600 —
— 500 —
— 400 —
WORKER'S WAGE
IN EACH   — 300 —
COUNTRY = 100
— 200 —
PROFESSOR
— 100 —

Where the Russian professor is a giant...
He earns 8 times as much as average Russian factory worker.
Our professors make only 1½ times U.S. factory worker's pay.

Sources: CENTER FOR INTERNATIONAL STUDIES, M. I. T.;
NATIONAL EDUCATION ASSOCIATION;
McGRAW-HILL DEPARTMENT OF ECONOMICS.

Figure 9-1. A much used but deceptive type of pictorial chart. The Russian professor "earns 8 times as much as the average Russian factory worker." Accordingly, the symbol representing the professor was made 8 times as tall, but in so doing, the area was made 64 times as large and the volume 512 times as large. (From McGraw-Hill Publishing Co., Inc., "Who are Today's Capitalists?" New York and Other Cities: Advertisement appearing during the week of November 17 to November 24, 1957 in 17 newspapers.)

thermore, if the symbols were measured correctly, charts of this kind would be divested of their most dramatic feature and seldom, if ever, be used.

Figure 9-1 illustrates clearly the implications of these statements. The data on which the chart is based indicate that a professor in Russia "earns 8 times as much as (the) average Russian factory worker." Accordingly, the designer made the symbol of the factory worker 1 unit high and the symbol of the professor 8 units high. In so doing, the area of the taller symbol is made 64 times as large as the smaller one, and the volume (human beings are three-dimensional) 512 times as large. The differential in the size of the two symbols is truly spectacular, but incorrect and deceptive. If the symbols had been constructed correctly, either on the basis of area or of volume, the differentials in size would be relatively slight and undramatic. Also, it should be noted that in the case of the symbols of the American professor and factory worker, although the ratio of earnings is only 1.5:1, the symbol representing the professor is 2.25 times as large in area and 3.38 times as large in volume.[4]

Figure 9-2 is a more detailed example of a pictorial chart in which the size of the symbol is the determinant criterion. The various sized oil barrels have been designed to reflect the price of oil charged by OPEC countries from January 1970 to July 1979. The symbol for 1970, the smallest in the series, represents $1.80 per barrel compared to the largest symbol, which indicates an estimated price of $20.00 per barrel forecast for July 1979. It is obvious that the 1979

[4] For a more detailed discussion of these problems, see Calvin F. Schmid and Stanton E. Schmid, *Handbook of Graphic Presentation,* New York: John Wiley and Sons, 1979, pp. 221–224. Also see Calvin F. Schmid, "What Price Pictorial Charts?" *Estadistica, Journal of the Inter-American Statistical Institute,* **14** (50) (March 1956), 12–25.

[5] Theoretically, if the larger symbol on the chart were scaled precisely in relation to the smaller one, the corresponding figures would be 123 times in area and 1371 times in volume.

[6] Willard C. Brinton, *Graphic Methods for Presenting Facts,* New York: The Engineering Magazine Company, 1914, p. 39.

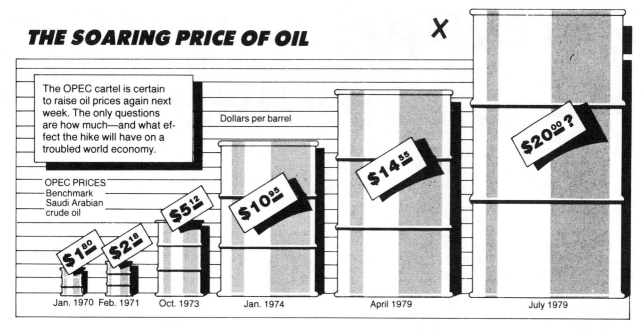

THE SOARING PRICE OF OIL

The OPEC cartel is certain to raise oil prices again next week. The only questions are how much—and what effect the hike will have on a troubled world economy.

Dollars per barrel

OPEC PRICES
Benchmark
Saudi Arabian
crude oil

X

$1⁸⁰

$2¹⁸

$5¹²

$10⁹⁵

$14⁵⁵

$20⁰⁰?

Jan. 1970 Feb. 1971 Oct. 1973 Jan. 1974 April 1979 July 1979

Figure 9-2. Another illustration of the type of pictorial chart in which the symbols ostensibly are made commensurate with the size or quantity reflected in the data. Almost invariably, however, only one dimension is used in gauging size or quantity. As a consequence, the size of the symbols are exaggerated. According to the data on the chart, the estimated price of oil per barrel in 1979 was 11.1 times as great as the price in 1970. But the 1979 symbol is close to 100 times as large in area and around 1000 times as large in volume. (From Newsweek, June 25, 1979, p. 65. Reproduced through courtesy of Bob Conrad. Copyright 1979, by Newsweek, Inc. All rights reserved. Reprinted by permission.)

Figure 9-3. For decades, graphic specialists have been preaching against the use of the "little man–big man" type of pictorial chart. In 1914, Willard C. Brinton included this illustration to substantiate his argument against the use of this type of chart. Also, from the basic data represented in the illustration, Brinton designed a pictorial unit chart. See Figure 9-4. (From Graphic Methods for Presenting Facts, by Willard C. Brinton. Copyright © 1914 McGraw-Hill Book Company. Used with the permission of McGraw-Hill Book Company.)

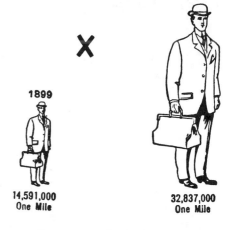

1911

1899

X

14,591,000
One Mile

32,837,000
One Mile

Passengers Carried on the Railroads of the United States in 1899 and in 1911 Compared

1899 14,591,000 ONE MILE

1911 32,837,000 ONE MILE

Number of Passengers Carried on the Railroads of the United States in 1899 and in 1911 Compared

Figure 9-4. An alternative to the "little man–big man" type of pictorial chart. As far as is known, this is the first pictorial unit chart to appear in print. Willard C. Brinton invented the pictorial unit chart in 1914. (From Graphics Methods for Presenting Facts, *by Willard C. Brinton. Copyright © 1914 McGraw-Hill Book Company. Used with the permission of McGraw-Hill Book Company.)*

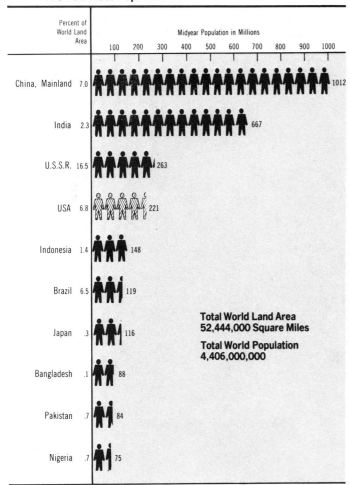

The Ten Most Populous Countries: 1979

Figure 9-5. A pictorial unit chart in which the basic graphic form is a simple bar chart. (From United States Bureau of the Census, Pocket Data Book U.S.A. 1979, *Washington, D.C.: Government Printing Office, 1980, p. iv.)*

price of $20.00 is a little over 11 times (11.1) as much as the price for 1970. However, the size of the 1979 symbol is close to 100 times as large in area as the 1970 symbol, and roughly 1000 times as large in volume.[5]

PICTORIAL UNIT CHART

For many decades specialists in statistical graphics have been well aware of the deficiencies of pictorial charts based on size and have inveighed against their use with much vehemence. In addition, attempts have been made to find an acceptable substitute. For example, in 1914, Willard C. Brinton comments that "Though this type of graphic work is common, it should be avoided, for its visual accuracy is serious enough to cause distrust of the whole graphic method."[6] As part of this discussion, Brinton reproduced an example of this type of chart that "gives the reader a false and exaggerated impression." (See Figure 9-3.) As a solution to this problem, he suggested a technique that has since become known as the "pictorial unit" type of chart. (See Figure 9-4.) He points out that Figures 9-3 and 9-4 are both drawn from the same data. "It was not a larger passenger, but more passengers, that railroads carried."[7]

Although Brinton was the first graphic specialist to formalize and recommend the pictorial unit chart, he never pursued its development and application further.[8]

The name of Otto Neurath is most commonly associated with the pictorial unit chart. Neurath invented it independently, entirely unaware of Brinton's work. Neurath's endeavors date back to 1923, when he first developed a series of pictorial charts for a housing exhibit in the city of Vienna under the auspices of the Austrian Housing and Garden Plot Association. In 1924, this exhibition was transferred into the famous Gesellschafts-und Wirtschaftsmuseum in Wien (Social and Economic Museum in Vienna).[9]

Besides creating an elaborate pictorial technique for portraying statistical and other data, Neurath later directed his energies toward the development of an auxiliary picture language—"A helping language," as he called it—"to be used along with other languages according to circumstances." The basic symbols of his system were called "isotypes" (International System Of Typographic Picture Education).[10]

In addition to Neurath himself, several of his former associates, particularly Rudolf Modley, have had a large part in the popularization and diffusion of the pictorial unit chart in the United States.

In designing a pictorial unit chart, full recognition should be given to certain basic requirements in the form of rules, principles, specifications and objectives. The first set of rules is relatively general as well as indicative of the purpose and character of pictorial unit charts:

1. Pictorial symbols should be self-explanatory. If the subject matter of the chart is ships, the symbols should clearly indicate ships.

2. Changes in numbers are shown by more or fewer symbols, not by larger or smaller ones.

3. Pictographs give only an overall picture; they do not show minute details.

4. Pictorial unit charts make comparisons, not flat statements.

5. Pictorial unit charts should be simple.[11]

Since the effectiveness of a pictorial unit chart depends so much on the symbol used, it is essential that symbols be appropriate and well designed. More specifically,

1. An artist should use the principles of good design established by fine and applied arts.

2. A symbol should be usable in either large or small size.

3. A symbol must represent a general concept, not a particular individual or species.

[7] Willard C. Brinton, *Graphic Methods for Presenting Facts,* New York: The Engineering Magazine Company, 1914, p. 39.

[8] Also, see Willard C. Brinton, *Graphic Presentation,* New York: Brinton Associates, 1939, pp. 121–131.

[9] Calvin F. Schmid and Stanton E. Schmid, *Handbook of Graphic Presentation,* New York. John Wiley & Sons, 1979, pp. 224–229.

[10] Marie Neurath and Robert S. Cohen, eds., *Otto Neurath Empiricism and Sociology,* Boston: D. Reidel Publishing Company, 1973, passim; J. A. Edwards and Michael Twyman, eds., *Graphic Communication Through Isotype,* Reading, England: University of Reading Press, 1975, passim; Otto Neurath, *International Picture Language,* London: K. Paul Trench, Trubner and Co., 1936, passim.

Home Values at Peak Levels in 1950

Median market values of all urban homes were **68** percent greater than for rural nonfarm homes. Values of white nonfarm homes were over **2½** times those of nonwhite homes.

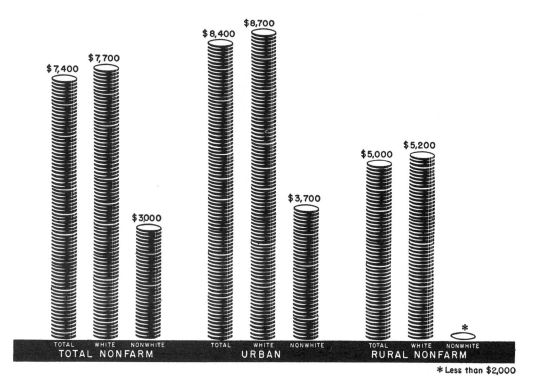

Figure 9-6.   The basic graphic design of this pictorial unit chart is a simple column chart. (From Housing and Home Finance Agency, **Housing of the Nonwhite Population, 1940 to 1950,** *Washington, D.C.: Government Printing Office, 1952, p. 6.)*

4. A symbol must be clearly distinguishable from every other symbol.

5. A symbol should be interesting.

6. A symbol is a counting unit and must be clear as such.

7. A symbol must be usable in outline as well as in silhouette.[12]

Another set of principles and procedures is concerned with the design and construction of the chart with the

various elements and other details put together in a unified and balanced manner. This implies that

1. The overall size and proportions of the chart should be singularly appropriate.

2. The choice of scale for the symbols should be compatible with the data and with the objectives of the chart.

3. The spacing of the symbols should be planned carefully so as to facilitate clarity, reliability, and harmony. This means that there should be an optimal number of symbols, not too close together and not too far apart. This also applies to the spacing of rows or columns of symbols.

4. The number of categories and the length of the rows or columns should be carefully determined.

[11] Rudolf Modley and Dyno Lowenstein, *Pictographs and Graphs: How to Make and Use Them,* New York: Harper and Brothers, 1952, pp. 24–28.

[12] Rudolf Modley and Dyno Lowenstein, *Pictographs and Graphs: How to Make and Use Them,* New York: Harper and Brothers, 1952, p. 47.

5. Fractions of symbols should be kept to a minimum. Sometimes it may be expedient to round off all figures in order to obviate the use of fractional symbols.

6. The style and size of lettering for the title, categories, notes, and legend should be chosen with discretion and positioned on the chart with extreme care.[13]

The following are examples of pictorial unit charts. The basic graphic form of Figure 9-5 is a simple bar chart. It portrays the populations of the world's 10 most populous countries in 1979. Each symbol represents 50 million people. Usually a legend or explanatory statement is included on the chart for the purpose of indicating the value of each symbol. However, it might be argued that in this instance a legend or statement would be superfluous since there is a scale, and the value of each category is clearly indicated by a figure at the end of each row of symbols.

The basic graphic form of Figure 9-6 is a simple column chart. The columns are represented by stacks of monetary symbols, each symbol valued at $100. As in the preceding chart, there is no statement indicating the specific value of each symbol. However, the value represented by each column is clearly indicated. Even if one were so inclined, it would be extremely troublesome, if not impracticable, to gauge the size of each column by actually counting the symbols.

The graphic form exhibited by Figure 9-7 is a bilateral bar chart. In fact, bars have been drawn as part of the chart with symbols representing $10 bills

_____

[13] Rudolf Modley and Dyno Lowenstein, *Pictographs and Graphs: How to Make and Use Them,* New York: Harper and Brothers, 1952, pp. 52–58.

Rents More Than Half Again as High as in 1940

Median gross rents have increased relatively more for rural nonfarm than for urban, and more for nonwhite than for white renters.

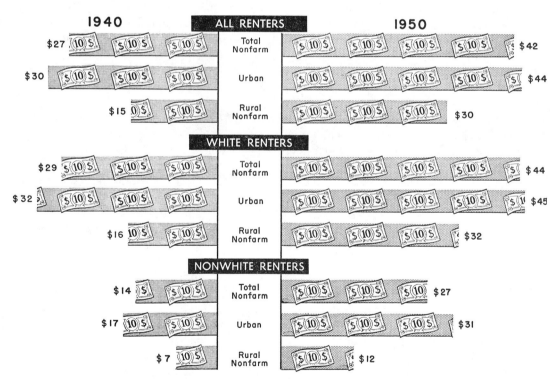

Figure 9-7. A bilateral bar design with pictorial units superimposed. (From Housing and Home Finance Agency, Housing of the Nonwhite Population, 1940 to 1950, Washington, D.C: Government Printing Office, 1952, p. 7.)

superimposed. Since the bars alone are clearly interpretable and have been incorporated as important
elements in the chart design, the monetary symbols
might be considered mere ornamentation. However,
the symbols seem to enhance the appeal of the chart
as well as serve as a readable and fairly reliable measuring gauge.

Figure 9-8 is a bilateral histogram with the X-axes
and Y-axes reversed. That is, contrary to usual practice, the class-intervals (age) are laid out vertically,
and the frequencies (percentages) are laid out horizontally. From the viewpoint of graphic design, the
symbols and general structure of the chart are excellent, but unfortunately with respect to statistical and
demographic convention and standards extending
over many decades, the chart does not conform to the
specifications of an authentic age-and-sex pyramid.
Rather, it can be described more properly as a
pseudo-age-and-sex pyramid. The statistical base in
the computation of percentages for an authentic age-
and-sex pyramid is the total population, not separate
male and female totals, as shown in Figure 9-8.

CHARTS IN WHICH THE BASIC GRAPHIC FORM IS PICTORIALIZED

This type of pictorial chart may assume many different forms, depending on the basic graphic design selected as well as the nature of the pictorialization.
Perhaps the most frequently used chart of this type is
the pie chart pictorialized as the traditional American
silver dollar. Examples of charts in which the silver
dollar representation is used will be found in our
Handbook of Graphic Presentation.[14]

Also, there are pictorialized dollar bills in the
form of a rectilinear coordinate line chart and a 100-
percent component bar chart. Other examples of pictorial charts of this kind included in the *Handbook*
are as follows: liquor barrels (pie charts showing proportions of regular drinkers, infrequent drinkers, and
abstainers), stone columns (column chart portraying
pronounced increase in governmental debt), and milk
can (component 100-percent column chart depicting
United States annual milk production).

---

[14] Calvin F. Schmid and Stanton E. Schmid, *Handbook of Graphic Presentation,* New York: John Wiley & Sons, 1979, pp. 225, 234, and 264.

Because of the varied form and structure that this
type of pictorial chart might assume, it is not possible
to list design standards and directions with any degree of specificity. Nevertheless, it is evident that the
basic prerequisites for designing effective charts of
this kind include a combination of expertise in statistical graphics and artistic skill. Figure 9-9 is a fairly
typical example of this kind of pictorial chart.

CHARTS WITH PICTORIAL EMBELLISHMENT

In the fourth type of pictorial chart, emphasis is
placed on various kinds of embellishments, primarily
for the purpose of stimulating interest or appeal. In a
secondary sense, pictorial embellishments may sometimes add to the interpretability of a chart. Characteristically, charts of this kind represent conventional
graphic forms embellished with pictorial symbols,
drawings, paintings, or photographs. Generally, pictorial symbols and sketches are superimposed on the
graphic form, while paintings and photographs more
commonly serve as a background for the graphic
form.

In designing a chart of this kind, certain minimal
requirements should be observed:

1. The primary graphic form should be designed
in accordance with acceptable standards.

2. The pictorial embellishments should be selected and executed with utmost discrimination
and expertise, which involves the following specific considerations:
(a) The pictorial representation should be appropriate to the basic design, message, data,
and purpose of the chart. (b) The size, proportions, and placement of the pictorial representations should be carefully planned. (c) From
an artistic point of view, the quality of the pictorial representation should meet the highest
standards of fine and applied arts.

Figure 9-10 is a simple surface chart with a photographic background. The photograph is descriptive of
the data depicted by the chart. The chart clearly reveals the decline in the number of wage workers in
agriculture. In a conventional, nonpictorial simple
surface chart, the area covered by the photograph
would have been stippled or crosshatched. The obvious question that comes to mind is, Which of the two

designs is better? A decision of this kind must be made within the context of the purpose or objective at hand as well as other factors.

Figure 9-11 is substantially more elaborate than the preceding chart, with particular emphasis on a fairly complex message. The message revolves around the American productivity crisis. The two causal factors that are emphasized in this process are the comparatively low rate of personal savings and the resultant low rate of investment. An examination of this chart reveals that the design reflects much care and ingenuity.

---

[15] W. H. Huggins and Doris R. Entwisle, *Iconic Communication: An Annotated Bibliography*, Baltimore: The Johns Hopkins University Press, 1974, p. 1.

Figure 9-8.   A pictorial unit chart in pyramidal configuration. Note the age and sex differentiation of symbols. The horizontal axis is expressed in percentages. It should be pointed out that two bases, the total number of males and the total number of females, were used in computing percentages for this chart. This is contrary to the well-established procedure of demographic experts in which the total population is used as the only base. Accordingly, this chart is not an authentic conventional age-and-sex pyramid. (From Rudolf Modley and Dyno Lowenstein, Pictographs and Graphs, How to Make and Use Them, *New York: Harper and Brothers, 1952, p. 44.)*

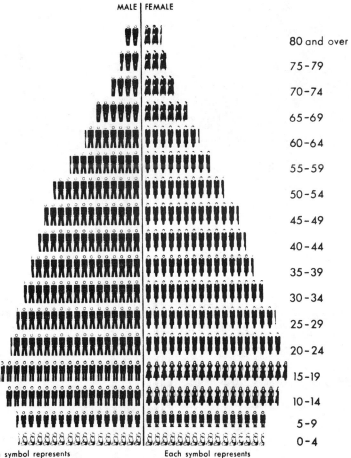

AGE DISTRIBUTION, 1940

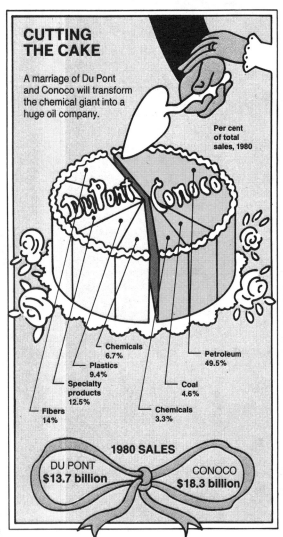

CUTTING THE CAKE

A marriage of Du Pont and Conoco will transform the chemical giant into a huge oil company.

Per cent of total sales, 1980

Chemicals 6.7%
Plastics 9.4%
Specialty products 12.5%
Fibers 14%

Petroleum 49.5%
Coal 4.6%
Chemicals 3.3%

1980 SALES

DU PONT $13.7 billion
CONOCO $18.3 billion

Figure 9-9. A chart in which the basic graphic form, a pie chart, it pictorialized as a wedding cake symbolizing the proposed merger of DuPont and Conoco. The sectors of the wedding cake indicate the major product areas of the two companies. (From Newsweek, July 20, 1981, p. 54. Copyright 1981 by Newsweek, Inc. All rights reserved. Reprinted by permission.)

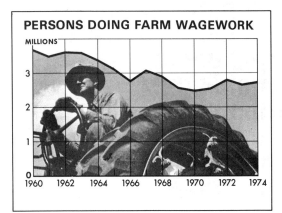

PERSONS DOING FARM WAGEWORK

MILLIONS

3
2
1
0

1960 1962 1964 1966 1968 1970 1972 1974

Figure 9-10. A rectilinear coordinate chart with a pictorially embellished background showing trends in farm wageworkers from 1960 to 1974. In a nonpictorial chart, the surface would be crosshatched or stippled. In this chart, the surface area is filled in with a photograph symbolizing a worker driving a tractor. (From United States Department of Agriculture, Handbook of Agricultural Charts, 1975, Agricultural Handbook No. 491, 1975, p. 57.)

CARTOON FACES: SHORT-LIVED FAD OR CONSTRUCTIVE INNOVATION?

In recent years some interest has been created by a special type of pictorial representation that was developed originally as part of an analytical statistical technique. Attempts have been made to adapt this innovation for the presentation and communication of statistical data. Since the emphasis in this book is on the process of communication rather than on analysis, this selection will present a brief exploratory and critical discussion of this method, particularly as it relates to communication.

This approach is linked in a general way with an area of interest referred to as "iconic communication." Briefly, "the word 'iconic,' from the Greek ikon, implies a mode of communication using primitive visual imagery that relies on the ability of people to perceive natural form, shape, and motion rather than on alphabetic symbols which are defined in terms of arbitrary conventions and which require special education to interpret."[15]

Included among the symbols used in graphic analytical techniques are "glyphs," "stars," "trees," "castles," and "cartoon faces." The symbols most frequently adapted for presentation and communication purposes are "cartoon faces" devised by Herman Chernoff.[16]

The original object of Chernoff's technique was "to represent multivariate data, subject to strong but possibly complex relationships, in such a way that an investigator can quickly comprehend relevant infor-

mation and then apply appropriate statistical analysis. . . . A sample of points in k-dimensional space is represented by a collection of faces."[17] The variables are encoded by and interpreted through a large number of physiognomic characteristics including the shape of the head, location and shape of the eyes and the eyebrows, curvature and width of the mouth, length of nose, shape of the upper and of the lower face. Significantly, electronic computer technology has made possible, in a practicable sense, the caricaturization of large numbers of cartoon faces.

A "graphic experiment in display of nine variables (that) uses faces to show properties of states" was attempted by Howard Wainer. (Figure 9-12.)[18] Wainer states that "We found that the use of Chernoff *faces* for this task yields an evocative and easy to understand display."[19] The display for Figure 9-12 was formed through the following variable/feature representations:

Population = > number of faces/state.

Literacy rate = > size of the eyes (bigger = higher).

Percent of high school graduates = > slant of the eyes (the more oriental, the better).

Life expectancy = > length of the mouth (the longer, the better)

Homicide rate = > width of the nose (the wider the nose, the lower the homicide rate).

Income = > curvature of the mouth (the bigger the smile, the higher the income).

Temperature = > shape of the face (the more like a peanut, the warmer; the more like a football, the colder).

Longitude and latitude = > x and y position of the face on the coordinate axes of the paper represent the position of the state.

This scheme is based on two assumptions. The first assumption, which is related to the encoding process, is that a particular variable can be reliably represented by the size, shape, or orientation of a cartoon face. The second assumption, which is related to interpretation, is that the "reader" possesses the ability to perceive and remember even small variations in the structure of human faces as portrayed by cartoon sketches. Accordingly, Figure 9-12 is supposed to reveal that "Angoff and Mencken's conclusion that Mississippi is the 'Worst American State,' followed

---

[16] Herman Chernoff, "The Use of Faces to Represent Points in k-Dimensional Space Graphically," *Journal of the American Statistical Association,* **68** (1973), 361–368; Herman Chernoff and M. Haseeb Rizvi, "Effect on Classification Error of Random Permutations of Features in Representing Multivariate Data by Faces," *Journal of the American Statistical Association,* **70** (1975), 548–554. Also, in this connection, see B. Kleiner and J. A. Hartigan "Representing Points in Many Dimensions by Trees and Castles," along with critiques by Howard Egeth, Robert J. K. Jacob, and Howard Wainer, *Journal of the American Statistical Association,* **76** (1981), 260–276. It may be of some interest to note that more recently a "new face . . . with 36 instead of 18 representable variables" has been proposed for the portrayal of multivariate data. See Bernhard Flury and Hans Riedwyl, "Graphical Representation of Multivariate Data by Means of Asymmetrical Faces," *Journal of the American Statistical Association,* **76** (1981), 757–765.

[17] Herman Chernoff, "The Use of Faces to Represent Points in k-Dimensional Space Graphically," *Journal of the American Statistical Association,* **68** (1973), 361–368.

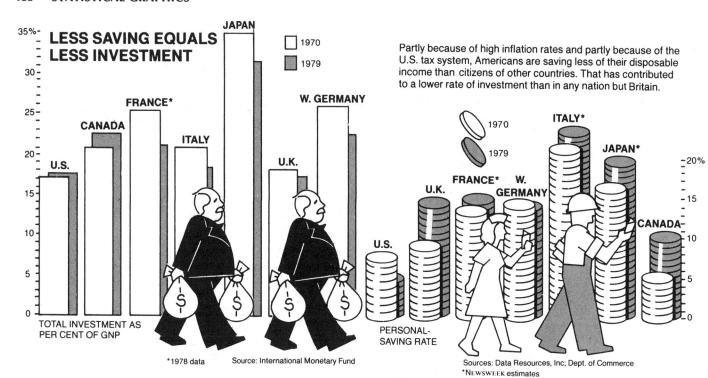

Figure 9-11. *An informative and well-planned chart with interesting and pertinent pictorial fillers illustrative of the chart's basic message. Even without a clear and comprehensive perception of all the factors contributing to the current American productivity crisis, there are two significant ones clearly depicted in this chart: low rate of personal savings and consequent low rate of investment. The ingenuity and ready interpretability manifested by this chart would give it a superior rating. (From Newsweek, September 8, 1980, p. 53. Reproduced through courtesy of Bob Conrad. Copyright 1980 by Newsweek, Inc. All rights reserved. Reprinted with permission.)*

closely by Alabama, South Carolina, Georgia, Arkansas, Tennessee, North Carolina, Louisiana, is still valid almost half a century later."[20]

In light of this experiment, it seems that the utili-

zation of the Chernoff cartoons as a technique of visual communication leaves much to be desired.[21] An effective chart at least should be readily interpretable and easily understood. These qualities as well as others are lacking in Figure 9-12. The deciphering of the facial characteristics of the cartoons, along with an attempt to derive a clear overall understanding of what the map purportedly conveys is a time-consuming and frustrating experience. Figure 9-12 seems to possess more the characteristics of an esoteric puzzle

---

[18] Howard Wainer, "Graphic Experiment in Display of Nine Variables Uses Faces to Show Properties of States," *The Newsletter of the Bureau of Social Science Research,* **XIII** (Fall 1979), pp. 2–3.

[19] The inspiration for selecting the data for this map dates back to an article published in 1931 in the *American Mercury* entitled "The Worst State." The article was authored by C. Angoff and H. L. Mencken.

[20] Chernoff is well aware of the limited value of cartoon faces as a communication device. For example, he states: "Faces can be used to communicate information to a limited extent *after some training.* . . . Faces are of little use to illustrate or communicate unless the audience is specially trained in which case they can be of limited use." Herman Chernoff, "Graphical Representation as a Discipline," in Peter C. C. Wang, *Graphical Representation of Multivariate Data,* New York: Academic Press, 1978, pp. 1–11.

[21] Also, it should be recognized that from a methodological point of view, answers to the basic statistical questions posed in this experiment could have been answered with greater precision, thoroughness, reliability, and expertise, if more appropriate and refined methods had been used. Of course, the primary purpose of the Wainer paper pertains to an experiment in statistical graphics, not statistical or ecological analysis. There is an extensive literature on the subject of areal analysis, from the point of view of both sociology and geography.

than a clear, straightforward, readily understood, and reliable vehicle of visual communication. In addition, it should be pointed out that the structure of an effective chart represents a well-organized visual composition in which the various elements are integrated into a clear, meaningful, and balanced whole. Important as symbols and other elements may be, they do not exist as independent and disparate attachments to a graphic display. Rather, in a well-designed chart, they are essential and integral parts of a total visual composition that is capable of conveying a message with clarity and reliability.[22]

Originality, ingenuity, and novelty have their place in statistical graphics, but unless the resultant design is clear and easy to understand, the chart becomes irrelevant or even worse, a serious liability. Statistical charts are not mere ornamentation, arcane

---

[22] In this connection, many empirical studies of symbols and other elements of statistical charts are of little value, because they have been analyzed as disparate elements, entirely divorced from the total setting in which they are supposed to function. For example, see Robert J. K. Jacob, "Symbols for Display of Multivariate Data: The Face," in *Proceedings of the First General Conference on Social Graphics,* Leesburg, Virginia, 1978, Technical Paper No. 49, United States Bureau of the Census, 1980, pp. 114–148.

Figure 9-12. *Application of cartoon faces for portraying the geographical distribution of several statistical indexes. See text for additional explanations and comments. (From Howard Wainer, "Graphic Experiment in Display of Nine Variables Uses Faces to Show Properties of States," The Newsletter of the Bureau of Social Science Research, XIII (Fall 1979).*

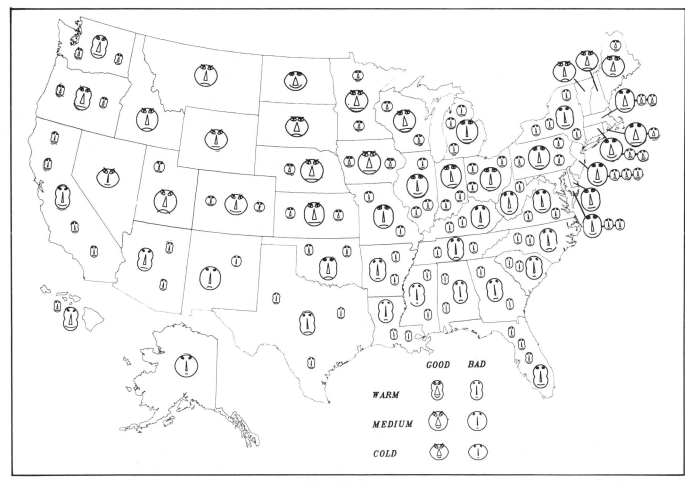

FACING THE NATION

Chernoff Faces of all fifty states showing how this meta-iconic scheme allows visual clustering in multidimensional space.

picture puzzles, or perfunctory adjuncts. Rather, in any well-ordered presentation, they are necessary and integral parts of the mainstream of communication. Effective communication demands the right chart in the right place. In a real sense, a statistical chart is a graphic essay, through the "eye–brain system," it conveys a message, not infrequently one that is complex, comprehensive, and detailed. As a vehicle of communication, a well-designed chart should be readily interpretable, accurate, reliable, uncluttered, and appealing, and no matter what form it may take, a chart should possess an optimal conceptual and physical structure. An acceptable chart is a composite of interrelated elements carefully designed to convey a specified message and adapted to the interests and abilities of users.

PRESENTATION OF ERRORS IN DATA

A Neglected Problem in Statistical Graphics

AN IMPORTANT BUT LONG-NEGLECTED problem in statistical graphics is concerned with appropriate design techniques for indicating with clarity and precision the reliability of data that are being graphically portrayed. Characteristically, these techniques represent special elements or components integrated with the chart selected for portraying the data.

To say that errors in statistical data are a common occurrence would not be an exaggeration. For example, errors in statistical data derived from surveys and censuses can be attributed to a number of sources such as sampling variability, response variability, response bias, nonresponse, falsification, and processing of data. More simply, errors can be subsumed under the dichotomy of sampling, and nonsampling errors.[1]

In light of statistical theory and practice, sampling errors and nonsampling errors pose entirely different problems of evaluation and presentation. Generally, sampling errors can be dealt with in a more consistent, systematic, and quantitative manner. Moreover, in terms of statistical standards, it is common practice for the original compiler or analyst to inform the consumer of statistical data of its quality, whether in textual, tabular, or graphic form. The primary emphasis in this chapter is on sampling errors.

The presentation of statistical errors in graphic form represents fundamentally a design problem, but one especially difficult because of rigorous constraints and special requirements: First, a presentation of this kind requires essentially a design based on the combination of two charts, the conventional graphic form chosen as most appropriate for the problem at hand integrated with one or more diagrammatic elements or components to provide a clear, visual indicator of errors. Second, a requirement of this kind demands that the resultant design avoid undue complexity, clutter, and confusion. Third, the elements or components that are added for measuring error must be reasonably precise, readily interpretable, and compatible with conventional chart design. Fourth, because of differences in shape, composition, and complexity among conventional charts, the integration of one or more elements for measuring errors with certain graphic forms may be very difficult and possibly cumbersome.

---

[1] A discussion of errors in spatially-ordered data and their implications in the design and construction of choropleth maps will be found in Chapter 6.

Although not related specifically to graphic design, it is nevertheless a significant fact in addressing a problem of this kind that the overwhelming proportion of those who make or use statistical charts seem to be oblivious to sampling and other errors. This state of affairs is one that has persisted for a very long time in spite of the fact that so many persons at present possess at least a rudimentary understanding of sampling and other statistical procedures, as well as of the implications of errors in statistical data. Also, in the utilization of statistical data, frequently no distinction is made between sample data and data for an entire universe. The uncritical and indiscriminate use of statistical data in the construction of charts may be accounted for by: (1) lack of knowledge and training in statistical theory and methods; (2) carelessness and indifference concerning the quality and critical evaluation of data; and (3) failure of agencies or individuals responsible for the data to indicate their essential characteristics and reliability.

BASIC CONCEPTS AND PROCEDURES IN PRESENTING ERRORS IN DATA

In order to make more explicit the problem of errors in data as well as to clarify certain pertinent statistical concepts and principles, extended reference will be made in this section to two excellent practical and relevant papers prepared under the auspices of the

**Unemployment Among High School Graduates
16 to 24 Years Old, by Race: October 1975**

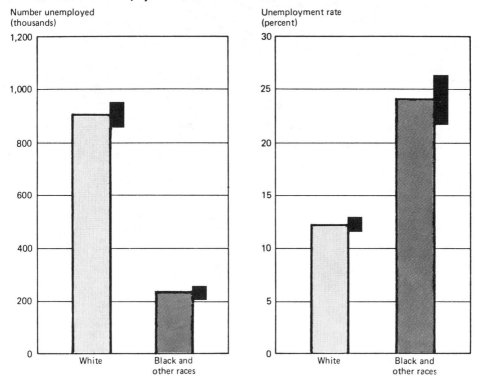

Figure 10-1.   The data depicted by this chart are based on a sample survey. The amount of sampling variability of the data is indicated by the small rectangles. The height of the rectangles covers a range of values of one standard error above and one standard error below the reported value. (From Office of Federal Statistical Policy and Standards, U.S. Bureau of the Census, Social Indicators, 1976, Washington, D.C.: Government Printing Office, 1977, p. XXVII.)

United States Bureau of the Census on *Standards for Discussion and Presentation of Errors in Survey and Census Data.*[2] These two papers embody basic standards along with supplementary explanations and illustrations concerning the presentation of errors in different types of publications and special tabulations. The Bureau of the Census and other federal

agencies consider it an important obligation to inform users of the basic methodological procedures utilized in collecting their data and of the errors and deficiencies that might occur.

The following discussion, extracted verbatim from United States Bureau of the Census Technical Paper 32 by Maria E. Gonzalez, Jack L. Ogus, Gary Shapiro, and Benjamin J. Tepping, is designed to clarify certain statistical concepts as well as to summarize briefly some of the more important problems involved in presenting errors in survey and census data.[3]

Sampling errors occur because observations are made only on a sample rather than on the entire population. Nonsampling errors can be attributed to many sources: inability to obtain information about all cases in the sample, definitional difficulties, differ-

---

[2] Maria E. Gonzalez, Jack L. Ogus, Gary Shapiro, and Benjamin J. Tepping, *Standards for Discussion and Presentation of Errors in Survey and Census Data,* Technical Paper 32, United States Bureau of the Census, Washington, D.C.: Government Printing Office, 1974. The second paper, which carries the same title, is essentially identical to Technical Paper 32, "but rearranged to better suit the needs of the more general user and patron of survey data." It was published in the *Journal of the American Statistical Association,* **70,** No. 351, Part II (1975), 1–23.

ences in the interpretation of questions, inability or unwillingness to provide correct information on the part of respondents, mistakes in recording or coding the data obtained, and other errors of collection, response, processing, coverage, and estimation for missing data. Nonsampling errors also occur in complete censuses. The "accuracy" of a survey result is determined by the joint effects of sampling and nonsampling errors.

The particular sample used in a survey is one of a large number of all possible samples of the same size that could have been selected using the same sample design. Estimates derived from the different samples would differ from each other. The deviation of a sample estimate from the average of all possible samples is called the sampling error. The standard error of a survey estimate is a measure of the variation among the estimates from the possible samples, and thus it is a measure of the precision with which an estimate from a particular sample approximates the average results of all possible samples. The relative standard error is defined as the standard error divided by the value being estimated.

The sampling estimate and an estimate of its standard error permit us to construct interval estimates with prescribed confidence that the interval includes the average result of all possible samples.

To illustrate, if all possible samples were selected,

Figure 10-2.   This chart, like the previous one, shows the range of 68-percent confidence limits of the data, not by separate rectangles but by cross-hatching. (From Maria E. Gonzalez et al., Standards For Discussion and Presentation of Errors in Data, Technical Paper 32, United States Bureau of the Census, Washington, D.C.: Government Printing Office, 1974, p. v-3.)

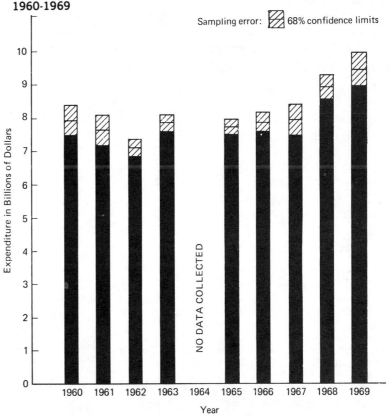

Estimates of Expenditures by Owner-Occupants for Alterations and Repairs to Their 1-to-4-Unit Residential Properties as Measured by Quarterly Surveys, 1960-1969

Note: Estimates for 1960 and 1961 are revised to adjust for bias in originally published figures, which we were able to measure and correct.

each of these were surveyed under essentially the same conditions, and an estimate and its estimated standard error were calculated from each sample. Then

1.   Approximately two thirds of the intervals from one standard error below the estimate to one standard error above the estimate would include the average value of all possible samples. We call an interval from one standard error below the estimate to one standard error above the estimate a two-thirds confidence interval.

2.   Approximately nine tenths of the intervals from 1.6 standard errors below the estimate to 1.6 standard errors above the estimate would include the average value of all possible samples. We call an interval from 1.6 standard errors below the estimate to 1.6 standard errors above the estimate a 90-percent confidence interval.

3.   Approximately nineteen twentieths of the intervals from two standard errors below the estimate to two standard errors above the estimate would include the average value of all possible samples. We call an interval from two standard errors below the estimate to two standard errors above the estimate a 95-percent confidence interval.

4.   Almost all intervals from three standard errors below the sample estimate to three standard errors above the sample estimate would include the average value of all possible samples.

Thus, for a *particular* sample, one can say with a specified confidence that the average of all possible samples, is included in the constructed interval.

Standard errors of estimated numbers of persons

---

[3] In addition to the two papers by Maria E. Gonzales et al. cited in Reference 2, the 100th edition of *Statistical Abstract of the United States, 1979,* United States Bureau of the Census, Washington, D.C.: Government Printing Office, 1979, contains a relatively brief but excellent review of "Statistical Methodology and Reliability." (See pp. 945–963.) This discussion covers a wide variety of data derived from a large number of sources including Bureau of Labor Statistics, Social Security Administration, Law Enforcement Assistance Administration, Internal Revenue Service, Bureau of the Census, Department of Agriculture, National Center for Education Statistics, and National Center for Health Statistics. Collection techniques, sampling and nonsampling errors, and other characteristics of the data are evaluated on a case-by-case basis.

Table 10-1

Absolute and Relative Standard Errors of Estimated Numbers of Persons

Size of Estimate	Standard Error	
	Absolute	Relative (%)
25,000	700	3.0
50,000	1,100	2.1
100,000	1,500	1.5
250,000	2,300	.9
500,000	3,300	.7
1,000,000	4,700	.5
2,500,000	7,400	.3
5,000,000	10,000	.2
10,000,000	15,000	.2
25,000,000	23,000	.1
50,000,000	33,000	.1

Table 10-2

Standard Errors in Percentage Points of Estimated Percentages

Estimated Percentage	Base of Estimated Percentage (thousands)			
	250	500	1,000	2,500
2 or 98	1.0	0.7	0.5	0.3
5 or 95	1.5	1.1	0.8	0.5
10 or 90	2.1	1.5	1.0	0.7
25 to 75	3.1	2.2	1.5	1.0
50	3.6	2.5	1.8	1.1
	5,000	10,000	25,000	50,000
2 or 98	0.2	0.1	0.08	0.06
5 or 95	0.4	0.3	0.1	0.08
10 or 90	0.5	0.4	0.2	0.1
25 or 75	0.7	0.5	0.3	0.2
50	0.8	0.6	0.4	0.3

may be read from Table 10-1, and standard errors of estimated percentages may be read from Table 10-2.

Table 10-1 gives both absolute and standard errors of estimates of the number of persons having any given characteristic covered by this report.

The standard error of a percentage, computed by using sample estimates for the numerator and denominator of the percentage, depends on the magnitude of the percentage as well as on the magnitude of the base of the percentage. The stub of Table 10-2 contains several levels of the estimated percentage.

Note that the standard error of an estimated percentage is the same as the standard error of the complement of the percentage. The column headings of Table 10-2 give several possible magnitudes of the base of a percentage, ranging from 250,000 persons in the population to 50,000,000 persons in the population. For example, if the percentage of the labor force that is unemployed is 5 percent and the number of persons in the labor force in the tabulation areas is 5,000,000, then the standard error of the unemployment rate is approximately 0.4 percentage points.

Standard errors for values within the ranges of the tables may be approximated by interpolation. We give first an example of interpolation in Table 10-1. The estimate of the number of employees of a certain class is 66,000, but the table shows standard errors for 50,000 and 100,000. The standard error is approximately

$$1100 + (1500 - 1100) \times \frac{66,000 - 50,000}{100,000 - 50,000} = 1228$$

which we should round to 1200. Interpolation in Table 10-2 may require interpolation for both the estimated percentage and the base for the estimated percentage. For example, the estimate for the proportion having a certain characteristic is 23.2 percent, on a base of 700,000. Interpolating between 10 and 25 percent for a base of 500,000, we obtain a standard error of 2.1 percentage points. Interpolating similarly for a base of 1,000,000, we obtain a standard error of 1.4 percentage points. Now interpolating between the bases of 500,000 and 1,000,000, we obtain a standard error of about 1.8 percentage points for a base of 700,000.

The following examples illustrate the use of standard errors in forming confidence intervals.

Figure 10-3. *Additional designs for portraying confidence limits for simple column and bar charts. Certain features based on suggestions of Albert Biderman and staff, Bureau of Social Science Research, Washington, D.C.*

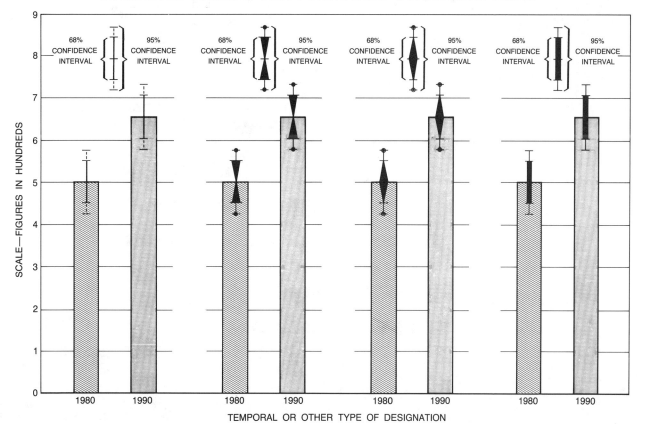

SUGGESTED DESIGNS FOR PORTRAYING CONFIDENCE LIMITS

Figure 10-4. The black bars on this chart have been designed to indicate 99-percent confidence intervals. This range is almost certain to include the estimate averaged over all possible repetitions of the sample. [Redrawn from Maria E. Gonzalez et al., "Standards for Discussion and Presentation of Errors in Data," Journal of the American Statistical Association, 70 (351), Part II (1975), 1–23.]

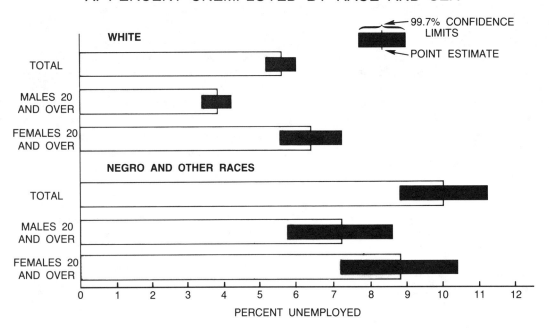

A. PERCENT UNEMPLOYED BY RACE AND SEX

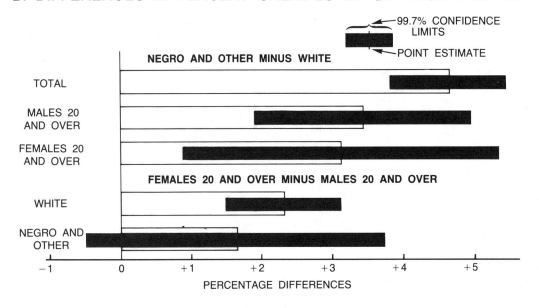

B. DIFFERENCES IN PERCENT UNEMPLOYED BY RACE AND SEX

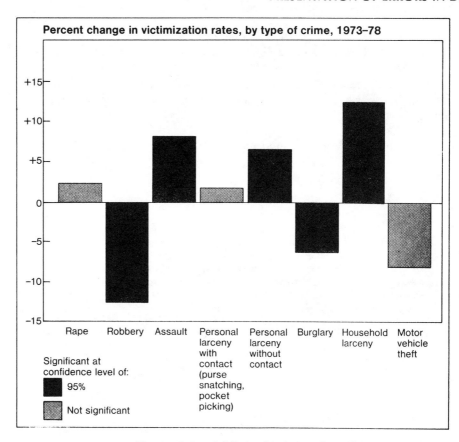

Figure 10-5. *A bilateral column chart showing percentage change in victimization rates between 1973 and 1978. The relative reliability of the percentages is shown by a shading system. The five percentages that are significant at the 95-percent confidence level are represented by black columns, while the remaining three percentages that are nonsignificant are shown by a lighter shading. [From U.S. Department of Justice, Bureau of Justice Statistics,* Criminal Victimization in the U.S.: 1973–78 Trends, *A National Crime Survey Report, NC5–N–13, NCJ–66716, December 1980, p. 3.]*

1. The estimate of the number of employees of a certain class is 66,000. Its estimated standard error was seen above to be 1200 employees. Based on these data, the two-thirds confidence interval is from 64,800 to 67,200 employees, and a conclusion that the average estimate of total employment lies within a range computed in this way would be correct for roughly two thirds of all possible samples. Similarly we could conclude that the average estimate of total employment derived from all possible samples, lies within the interval from 64,-100 to 67,900 employees with 90-percent confidence, that the average estimate of total employment lies within the interval from 63,600 to 68,400 employees with 95-percent confidence, and that the average estimate of total employment lies within the interval from 62,400 to 69,600 employees almost certainly.

2. The estimate of the proportion having a certain characteristic is 23.2 percent on a base of 700,-000, and its estimated standard error was seen above to be 1.8 percentage points. Based on these data, the two-thirds confidence interval for this sample is from 21.4 to 25.0 percent. A conclusion that the average estimate of the percentage derived from all possible samples lies within a range computed in this way would be correct for roughly two thirds of all possible samples. Similarly, we can conclude that the average estimate of the percentage derived from all possible samples lies within the interval from 20.3 to 26.1 percent with 90-per-

cent confidence in our conclusion, or that the average estimate of the percentage lies within the interval from 19.6 to 26.8 percent with 95-percent confidence, or that the average estimate of the percentage lies within the interval from 17.8 to 28.6 percent almost certainly.

3.   The estimate of a certain characteristic for the North is 23.2 percent and the estimate of the same characteristic for the South is 20.0 percent. Since both of these estimates are based on samples, the estimated difference of 3.2 percentage points (23.2 − 20.0 = 3.2) is also subject to sampling error. If the base of both 23.2 and 20.0 percent were 700,-000, Table 10-2 shows that the estimated standard error of 23.2 percent is 1.8 percentage points, and the estimated standard of 20.0 percent is 1.7 percentage points. A rough estimate of the standard error of the absolute standard errors can sometimes be related to the estimate more easily than

Figure 10-6. A rectilinear coordinate line chart with four curves based in part on sample data. The 95-percent confidence levels for the respective series of annual rates from 1954 to 1967 are shown by vertical lines, with the upper and lower limits marked with short horizontal lines. (From Abbott L. Ferriss, Indicators of Change in the American Family, New York: Russell Sage Foundation, 1970, p. 15. Copyright 1970 by Russell Sage Foundation. By permission of Russell Sage Foundation.)

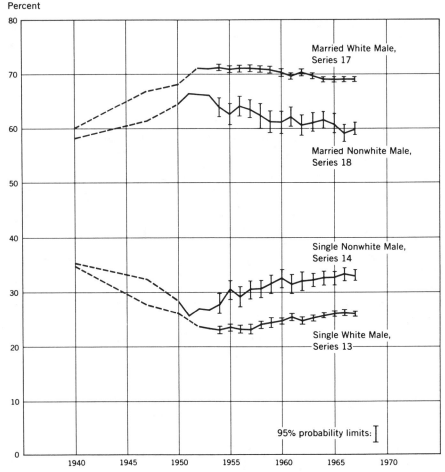

Marital Status of Males, 14 Years of Age and Over, by Color, 1940−1967, Series 13−14 and 17−18, and the Standard Error of the Estimated Percentages, 1954−1967

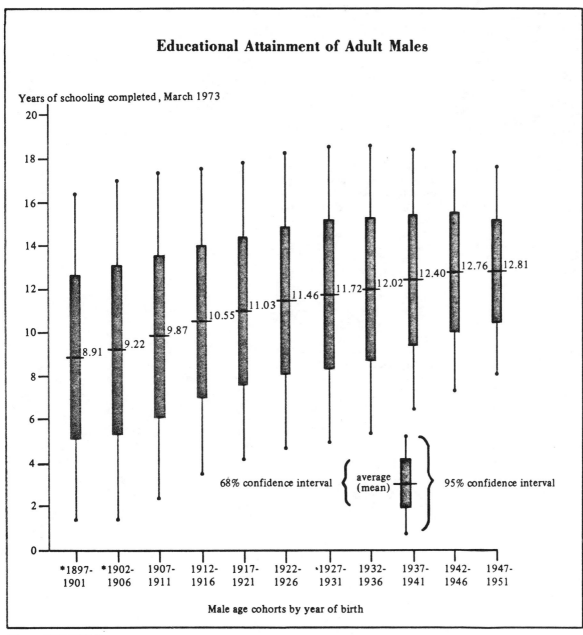

Figure 10-7. In this chart, the portrayal of sampling variability of the basic data has been incorporated as an integral part of the chart design. The vertical bars and lines indicate specifically 68- and 95-percent confidence intervals, while short horizontal lines and figures show estimates for each time period. (From Mary A. Golladay, The Condition of Education, Part 1, HEW National Center for Education Statistics, Washington, D.C.: Government Printing Office, 1977, p. 104.)

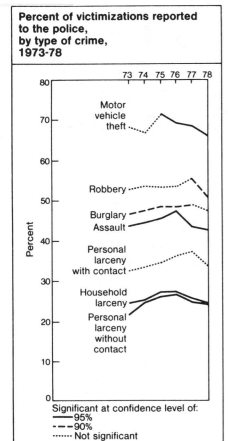

Percent of victimizations reported to the police, by type of crime, 1973-78

Significant at confidence level of:
——— 95%
– – – 90%
········ Not significant

Figure 10-8. Another rectilinear coordinate line chart in which the reliability of the data are indicated by curve patterns. The basic design principle of this chart is excellent, but its construction features are of mediocre quality. [*From U.S. Department of Justice, Bureau of Justice Statistics, Criminal Victimization in the U.S.: 1973–78 Trends, A National Crime Survey Report, NCS–N–13, NCJ–66716, December 1980, p. 6.*]

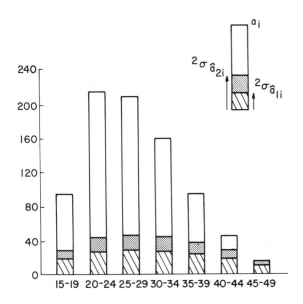

Figure 10-9. Hypothetical age-specific fertility rates (a_i) and measures of two standard deviations of estimated rates as derived from recent births ($2\sigma_{\hat{a}_{1i}}$) and current pregnancies ($2\sigma_{\hat{a}_{2i}}$). The two standard deviations represent a 95-percent confidence interval [*From Noreen Goldman and Charles F. Westoff, "Can Fertility be Estimated from Current Pregnancy Data?" Population Studies, 34 (1980), 535–550.*]

relative standard errors. This is especially apt to be the case for estimated percentages. In some instances, however, the relative error is of greater interest. Relative standard errors may require less space and may be applicable to a larger number of estimates in a given table, especially for aggregates that vary greatly in size or in their unit of measure. The preference between absolute standard errors and relative standard errors will vary depending on the nature of the table. The choice should be made by the authors in each individual case. The importance of avoiding ambiguity when presenting absolute standard errors of percentages is evident. The need to distinguish clearly between the absolute number of percentage points and the concept of the relative error, in percent, is clearly recognized. For example, if the standard error of an estimated 38.8 percent is .4 percentage points, this would represent a relative standard error of approximately 1 percent.[4]

ILLUSTRATIONS OF DESIGN TECHNIQUES FOR PRESENTING SAMPLING ERRORS

Figure 10-1 is a simple column chart that compares, according to race, the number and percentage of unemployed persons 16 to 24 years old who completed four years of high school. The data represented by this chart are based on a sample compiled in the Current Population Survey of October 1975. The small rectangles on the chart denote the range of values obtained by adding and subtracting one standard error to the reported value. For the white category, the civilian labor force was 7,417,000; the number unemployed, 909,000; percentage unemployed, 12.25; approximate standard error for the number of unemployed, 41,000; and approximate standard error for percentage unemployed, 0.55. The corresponding figures for black and other races: labor force, 962,000; number of unemployed 232,000; percentage unemployed 24.12; standard error for number unemployed 22,000; and standard error for percentage unemployed, 2.28. The reported value, plus and minus the standard error, defines the range of values. The chances are about 68 out of 100 that the true value that would have been obtained for a complete count will fall within that range. Moreover, it will be recalled from the preceding discussion that an interval from two standard errors below the estimate to two standard errors above the estimate represents a 95-

percent confidence interval. This signifies that the chances are about 95 out of 100 that the true value (based on a complete census) will fall somewhere within this wider range. The larger relative standard error for the "black and other races" category is primarily a function of smaller sample size.[5]

Figure 10-2 represents another design technique that has been used to indicate the reliability of estimates for a simple column chart. It will be seen that the hatched portions of the several columns include the range of one standard error above and one below the sample survey estimate. It will be recalled that the chances are about 68 out of 100 that an estimate from the sample would differ from the value obtained in a complete census by less than the plus-minus standard error range. The data presented by this chart shows estimates of expenditures by owner-occupants for alterations and repairs to their 1-to-4 family properties. The expenditures indicated on the vertical scale are expressed in billions of dollars.[6]

In addition to the two design techniques illustrated by Figures 10-1 and 10-2 for portraying confidence limits on simple column charts, other techniques may be used. These techniques are characteristically in the form of one or more lines or symbols extending vertically from inside the upper portion of a column to outside the column, with appropriate horizontal lines or other markings indicating confidence interval limits. Figure 10-3 embodies a few suggested illustrations of these designs. It is apparent that the several techniques for portraying confidence limits on simple column charts can also be readily adapted to bar charts.

Figure 10-4 is an illustration of a bar chart based on sample data along with an indication of associated 99-percent confidence intervals. The data pertain to

---

[4] This extended quotation taken from Maria E. Gonzalez, Jack L. Ogus, Gary Shapiro, and Benjamin J. Tepping, *Standards for Discussion and Presentation of Errors in Data,* Technical Paper 32, United States Bureau of the Census, Washington, D.C.: Government Printing Office, 1974, Appendix 1, pp. 1-1–1-3.

[5] Office of Federal Statistical Policy and Standards, U.S. Bureau of the Census, *Social Indicators, 1976,* Washington, D.C.: Government Printing Office, 1977, pp. XXVI–XXVII.

[6] Maria E. Gonzalez, Jack L. Ogus, Gary Shapiro, and Benjamin T. Tepping, *Standards for Discussion and Presentation of Errors in Data,* Technical Paper 32, United States Bureau of the Census, Washington, D.C.: Government Printing Office, 1974, Appendix V, p. 3.

ESTIMATES OF U.S. ULTIMATE OIL PRODUCTION BASED ON DISCOVERY DATA

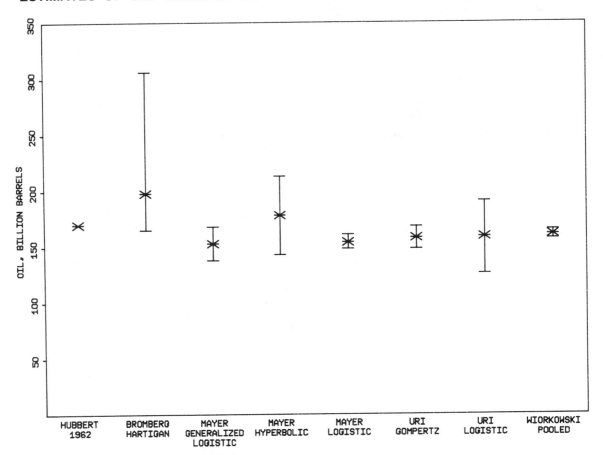

Offshore and Alaska are excluded.
The horizontal bars define approximate 95 percent confidence intervals.
The point estimate is denoted by an asterisk.

Figure 10-10.   The point estimates on this chart are represented by asterisks and the 95-percent confidence intervals by vertical lines. The chart depicts eight estimates of ultimate oil production in the United States. Except for the title and explanatory notes, this chart was constructed on an electronic computer. [From John H. Schuen-emeyer, "Comment on Estimating Fuel Resources," Journal of the American Statistical Association," 76 (September 1981), 554–558.]

unemployment rates by race, sex, and age. The black bars in the upper portion of the chart (A) cover a confidence range that is virtually certain to include the estimate averaged over all possible repetitions of the sample. Similarly, in the lower portion of the chart (B), the black bars provide the same information for estimated differences in unemployment rates among the various groups. For example, from (A) it can be stated with nearly certain confidence that the unemployment rate for white males 20 years of age and over lies between 3.4 and 4.2 percent. For males of Negro and other races 20 years and over, the comparable range is 5.8 to 8.6 percent. The lower portion

(B) of Figure 10-4 provides more specific information about the estimated differences among these classes. With nearly certain confidence, it can be stated that the unemployment rate for males 20 years or over of Negro and other races is 1.9 to 4.9 percentage points higher than for white males 20 and over.[7]

Figure 10-5 illustrates another method for portraying graphically the comparative reliability of a se-

---

[7] Maria E. Gonzalez, et al., "Standards for Discussion and Presentation of Errors in Data" *Journal of the American Statistical Association,* **70**, No. 351, Part II (1975), 10.

Performance of 13-Year-Olds and 17-Year-Olds on Career and Occupational Development Test

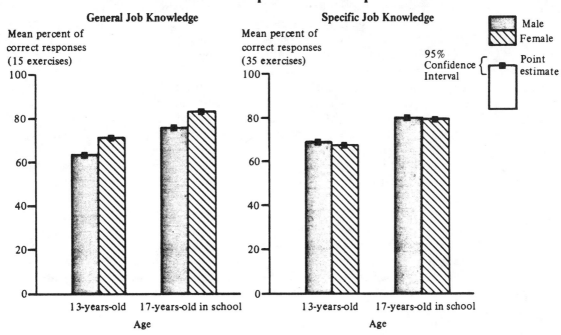

Figure 10-11. *In this chart, the point estimates are designated by small black squares at the top of each column and the 95-percent confidence intervals are measured by a bracketed distance shown in the legend.* [*From Mary A. Golladay,* The Condition of Education, *HEW National Center for Education Statistics, Washington, D.C.: Government Printing Office, 1977, p. 110.*]

ries of estimates by means of a shading scheme, with black columns representing estimates that are "significant at the confidence level of 95%" and lighter shaded columns "not significant" at that confidence level. The basic design of Figure 10-5 is a bilateral column chart representing increases (above the base line) and decreases (below the base line). Of course, the basic design could have been a bar chart if the delineator had arranged the base line vertically rather than horizontally. The chart portrays percentage increases and decreases in victimization rates by type of offense between 1973 and 1978.[8]

Figure 10-6 is an example of a rectilinear coordinate chart that includes as part of its design an indication of the reliability of the data represented by the more recent portions of the four curves. Portions of

the four curves based on annual sample estimates extend from 1952 to 1967. The criterion of reliability chosen for this chart is the 95-percent confidence interval. It is symbolized on the curves at yearly intervals by vertical lines with small horizontal lines at each end apparently to add emphasis to the confidence limits.

Another temporal series of estimates derived from sample studies in which 68 percent and 95 percent confidence intervals are superimposed are depicted in Figure 10-7. The estimates of educational attainment for each of the male age cohorts are arithmetic averages (means) of school years completed. The average number of years of schooling completed for adult males varies from 8.91 years for the earliest age cohort, 1897–1901, to 12.81 years for the most recent age cohort, 1947–1951.

Figure 10-8 depicts trends in the yearly percentage of victimizations reported to the police by type of crime from 1973 to 1978, including a clear indication of the reliability of the data. The levels of significance

---

[8] United States Department of Justice, Bureau of Justice Statistics, *Criminal Victimization in the U.S.: 1973–78 Trends,* A National Crime Survey Report, NC5-N-13, NCJ-66716 (December 1980), p. 3.

of the seven series of percentages are differentiated by three curve patterns. Data significant at the 95 percent confidence level are shown by a full curve, data significant at the 90 percent level by a dashed curve, and data not significant at either level by a light dotted curve. Incidentally, from a graphic standpoint the design of Figure 10-8 is mediocre.

In Figure 10-9 a series of hypothetical fertility rates are depicted together with a measure of two standard deviations of estimates (95 percent confidence interval) based on recent births ($\sigma_{\hat{a}_{1i}}$) and current pregnancies ($\sigma_{\hat{a}_{2i}}$). It is apparent from this chart that sampling errors involved in estimating age-specific fertility rates are often relatively large, especially for older age groups.

Figure 10-10 is a graphic summary of eight estimates of ultimate oil production in the United States based on studies in which various models and statistical estimation techniques were used. These studies are identified on the chart by names of the several investigators. Where error data are available, 95-per-cent confidence intervals have been plotted on the chart. The technique used in portraying this information is very simple, with point estimates designated by asterisks and the 95-percent confidence intervals by vertical lines of varying length. It will be seen that the graphic technique in Figure 10-10 is similar to that displayed in other charts in this chapter except for the fact that Figure 10-10 does not use complete columns to represent point estimate values.

Figure 10-11 is a simple column chart that depicts comparisons of occupational test results according to sex and age. In basic design it is a relatively slight variation of other charts included in this chapter. The point estimate values are designated at the top of each column by a small black square, and the 95-percent confidence interval by a bracketed space in the legend, entirely separated from the eight columns in the chart. Incidentally, the chart indicates that females both 13 and 17 years old perform better than males on general job knowledge, but that on specific job knowledge both sexes are about the same.

NAME INDEX

SUBJECT INDEX